RADICAL

AMAZEMENT

RADICAL

AMAZEMENT

CONTEMPLATIVE LESSONS
FROM BLACK HOLES, SUPERNOVAS,
AND OTHER WONDERS
OF THE UNIVERSE

JUDY CANNATO

SORIN BOOKS Notre Dame, Indiana

www.sorinbooks.com

ISBN-10 1-893732-99-1

ISBN-13 978-1-893732-99-5

Cover design by Angela Moody, Moving Images

Text design by Katherine Robinson Coleman

Printed and bound in the United States of America.

Library of Congress Cataloging-in-Publication Data
Cannato, Judy.
  Radical amazement : contemplative lessons from black holes, supernovas, and other wonders of the universe / Judy Cannato.
      p. cm.
  Includes bibliographical references.
  ISBN-13: 978-1-893732-99-5 (pbk.)
  ISBN-10: 1-893732-99-1 (pbk.)
  1. Religion and science. 2. Cosmology. I. Title.

BL240.3.C34 2006
231.7'652—dc22

2005027363

CONTENTS

FOR PHIL

There are many family, friends, and colleagues whom I want to thank for creating the morphogenic field in which this book unfolded. Bob Hamma has been a gentle mentor, full of good suggestions and welcome encouragement. Julie Cinninger, Mary Andrews, Christina Monroe, and Peter Gehred of Ave Maria Press have tirelessly supported my books with their creative gifts. My family continues to be of support: my husband Phil, who continues to be a partner in every writing project and to whom I dedicate *Radical Amazement*; our sons Phil and Doug, who share our joy; my mother Lucille, who is an inspiration; and my sister Linda, who calls to ask "Are you writing?" Carol Creek, Sr. Pat Kozak, CSJ, Sr. Jacquelyn Gusdane, SND, Sr. Carol Zinn, SSJ, Mary Ann O'Hara, Liz Vajentic, and Bob and Marie Rubey have been part of ongoing conversations and experiences that I find invaluable. Thank you!

Special recognition and thanks to the Congregation of St. Joseph of Cleveland, Ohio. I have been an associate member of the community for a decade, drawn by the charism of unity and reconciliation and inspired by their work. This experience of community has encouraged me to grow and use my gifts, sharing with the community in ministry and mission in a way that seeks to make the charism tangible. No words can adequately express my gratitude for their support and my love for who they are, a congregation of women who are without a doubt radically amazing! Sisters, I hope the charism shines through in these pages!

INTRODUCTION

Awareness of the divine begins with wonder.

∗ ABRAHAM HESCHEL

Thomas Aquinas said that a mistake in our understanding of creation will necessarily cause a mistake in our understanding of God.[1] Imagine what that means for us who live in an age in which scientific discoveries have taken us far beyond the truths we held in our youth. Our understanding of the universe has undergone a revolution in our lifetime, and if what Thomas says is true, we must accept the challenge to rethink and reshape our relationship to the divine in a way that resonates with the new discoveries about creation.

When I facilitate days of reflection with groups I often begin by placing a thimble full of sand on a large dark blue circle of paper and then inviting retreatants to enter into the following exercise:

∗ Imagine that each grain of sand is a star. On a very clear night, in the right location, with good eyes and no light pollution, we can see two to three thousand stars.

∗ Pick out one grain of sand to represent our Sun. Imagine our solar system with all its planets and planetoids orbiting as they have for about five billion years. Now look for the next closest grain of sand. That is Proxima Centauri, part of the Alpha Centauri system, only 4.35 light years away. Doesn't that sound close? Yet light travels at the speed of 186,000 miles per second, or 670 million miles per hour. A beam of light can travel from New York to Los Angeles in 0.016 seconds. It can pass around Earth at the equator in 0.133 seconds. The moon is 239,000 miles away, yet a light beam can get there in only 1.29 seconds and go beyond to the sun (93 million miles away) in just eight

7

minutes. Such incredible speed, yet it takes over four years to reach our closest neighboring star! If we went out tonight, looked into the dark sky and spotted Proxima Centauri, the light that reaches our eyes—traveling at 186,00 miles per second—will have left there 4.35 years ago.

* Now imagine that around the grain of sand that represents our Sun there is a coherent grouping of stars. This group represents the Milky Way, a spiral-shaped galaxy about one hundred thousand light years across and ten thousand light years deep, with between two hundred and four hundred billion stars. To continue the analogy with grains of sand, to represent all the stars in the Milky Way galaxy we would need a rather large dump truck.

* As recently as the 1920s we thought that the Milky Way galaxy comprised the entire universe, but in 1923 Edwin Hubble photographed the Andromeda galaxy which lies 2.5 million light years away. Today we know that there are billions of galaxies, each with billions of stars. The Hubble Space Telescope, launched in 1990 and orbiting 375 miles above Earth, has been able to capture images approximately twelve *billion* light years away!

* How many grains of sand does it take to represent the stars in the known universe? Try to imagine railroad hopper cars, filled and passing by at the rate of one per second, twenty-four hours a day, seven days a week. It would take three years for all the sand-filled cars to pass by![2] Our universe is immense, boggling the mind in any attempt to grasp its expanse.

* Beside discovering the existence of other galaxies, Edwin Hubble confirmed that the universe is expanding, a major revolution in the previous concept of a static, fixed cosmos that scientists had always envisioned. Even Einstein resisted the notion of an expanding universe.

RADICAL AMAZEMENT

* The discovery that the universe is expanding led the Belgian priest and physicist Georges Lemaître to assert that the universe must have had a starting point, just as the book of Genesis says. In 1949 George Gamow formulated the theory now known as the Big Bang. Although it has been modified along the way, this theory remains the best explanation of the origin of the universe to date.

* In 1998 Wendy Freedman and a team of astrophysicists concluded that the Big Bang occurred about 13.7 billion years ago.

* In 2003, using the Wilkinson Microwave Anisotropy Probe (WMAP), scientists determined that about twenty-five percent of the universe is what is called dark matter. Dark matter cannot be seen but can be detected by the gravitational pull it exerts. Even more amazing, approximately seventy percent of the universe is dark energy, the force that is causing the rate of expansion of the universe to accelerate. Only about five percent of the universe is composed of ordinary matter—what we can observe directly in some way. In other words, there is ninety-five percent more universe out there than what we can see!

* Finally, some physicists have come to the conclusion that from the first moment of the Big Bang there has been the impetus toward life. If the universe had unfolded *one trillionth of a trillionth* of a percent slower, the gravitational force would have been too great and the universe would have imploded before anything of significance occurred. If, however, the universe had unfolded one trillionth of a trillionth of a percent faster, matter would have escaped the gravitational pull and the cosmos would have been flung apart before anything could happen. It is not far-fetched to conclude that there has been an intentionality toward life all along.

The discoveries made by science over the last century have changed how we tell the story of creation. These findings provide new images for reflection, new metaphors for understanding. As the new universe story seeps into our awareness, it challenges us to expand the way we think about and respond to the life around us. As clearly as the parables told by Jesus challenged his listeners to ask questions about who they were and what their relationships meant, so the new universe story challenges us.

How are we to respond to the information and images that are becoming part of our everyday reality? May I suggest awe, wonder—*radical amazement*—as we encounter the Mystery that is everywhere around us? Nothing less is possible or appropriate than a radical amazement that brings us to shout or whisper "Oh my, God!" No matter what our religious convictions, it is evident that we live and move and have our being in the midst of a Mystery that is deeper than ourselves and broader than our own creativity and genius can possibly grasp.

The phrase "radical amazement" comes from Abraham Heschel, who said that wonder or radical amazement is the chief characteristic of a religious attitude toward life and the proper response to our experience of the divine. The insights that connect us to the Holy One come "not on the level of discursive thinking, but on the level of wonder and radical amazement, in the depth of awe, in our sensitivity to the mystery, in our awareness of the ineffable."[3] Living in radical amazement brings us into the space in which "great things happen to the soul."[4]

Although it may begin with an event that catches us up in awe and wonder, radical amazement goes beyond the event itself. According to Heschel, radical amazement "refers to all of reality; not only to what we see, but also to the very act of seeing as well as to our own selves, to the selves that we see and are amazed at their ability to see."[5]

The truth of Heschel's observation found a personal resonance last fall when I made a retreat at the mountain-top home of friends in North Carolina. The house sat at an elevation of over 5500 feet, and was far enough from any city to avoid the lights on Earth obscuring those in the heavens. On one of the evenings, long after sunset, I sat on the deck and turned my attention to the cloudless sky. There, splayed overhead in brilliant array, were the stars of the Milky Way. Not since childhood do I recall seeing with such clarity the magnificence of the night sky. The stars seemed to form a continuous carpet of light that trailed off into Mystery. I gasped involuntarily as I was swept up in awe. In that moment there was no distance between my body and the stars, no separation between myself and Mystery. This, undoubtedly, was an experience of radical amazement!

The amazement did not end when I went inside. I believe I took the moment—and the light of the stars themselves—within and now hold the encounter with all its delight and wonder in my heart. There I marvel not only over the capacity to see, but also for the gift of reflection, the ability to integrate each new vision with all previous experience, to know that I am connected to all that has been, all that is now, and all that is to come. I can choose to live in contemplative fidelity to each moment of radical amazement, knowing that something great is indeed happening to my soul.

All of life—every bit and particle of experience in every arena we inhabit and at every level we have awareness— invites us into the experience of radical amazement that is a doorway to the divine. "It is not the beginning of knowledge but an act that goes beyond knowledge; it does not come to an end when knowledge is acquired; it is an attitude that never ceases."[6] Radical amazement, then, is the stance in which we are invited to live, a cultivated way of life filled with attentiveness and vision.

In *The Silent Cry* German theologian Dorothee Soelle writes "I think that every discovery of the world plunges us into jubilation, a radical amazement that tears apart the veil of triviality."[7] When the veil is torn apart and our vision is clear there emerges the recognition that all life is connected—a truth not only revealed by modern science but resonant with ancient mystics. We are all one, connected and contained in a Holy Mystery about which, in all its ineffability, we cannot be indifferent.

Soelle maintains that radical amazement is the starting point for contemplation. Often we think of contemplation as a practice that belongs in the realm of the religious, some esoteric advanced stage of prayer that only the spiritually gifted possess. This is not the case. Although there are stages of prayer through which anyone who devotes time to prayer will normally progress, the nature of contemplation as I describe it here is one that lies well within the capacity of each of us. To use a familiar phrase, contemplation amounts to "taking a long loving look at the real."

We hardly take a long look at anything these days. In our day-to-day lives we often move at such a hurried pace that the best we can do is tender a brief glance. A friend recently told me about her experience with her grandniece. Not feeling attended to in the manner in which she was accustomed, the child, age three, climbed on my friend's lap, grabbed her face, and turned it toward her, saying "Wook at me!" Three-year-olds are naturally contemplative, knowing the importance of seeing what is in front of us and the distress that occurs when something radically amazing has just passed by unnoticed.

Contemplation also involves a loving look. Sometimes when we do pause long enough to engage what is around or in us, we do so with something less than love. Absent is a sense of wonder as we register disapproval of ourselves and others. Gone is compassion and mercy, replaced by an inclination to censure and judge. The contemplative stance that flows out of radical amazement catches us up in love—the Love that is the Creator of all that is, the Holy Mystery that never ceases to amaze, never ceases to lavish love in us, on us, around us.

Contemplation is a long loving look at what is real. How often we are fooled by what mimics the real. Indeed, we live in a culture that flaunts the phony and thrives on glittering fabrication. We are so bombarded by the superficial and the trivial that we can lose our bearings and give ourselves over to a way of living that drains us of our humanity. Seduced by the superficial, we lose the very freedom we think all our acquisitions will provide. When we are engaged in the experience and practice of radical amazement, we begin to distinguish between the genuine and the junk. Caught up in contemplative awareness and rooted in love, we begin to break free from cultural confines and embrace the truth that lies at the heart of all reality: We are one.

The invitation to be contemplative is nothing new, but it now carries with it an urgency particular to our time. This call to live contemplatively is offered to everyone. Often we want to relegate such a practice or lifestyle to the "religious" or "spiritual" in our midst, but the simple truth is that we have all been given eyes to see. We simply need to choose to live with vision. What is becoming more apparent by the day is that we must all become contemplatives, not merely in the way we reflect or pray, but in the way we live—awake, alert, engaged, ready to respond in love to the groanings of creation. Human life depends upon our living this way.

It is important not to confuse contemplative living with passive living. In our faith tradition, contemplation has always required response—not just interiorly, but out in the world where the fruits of our prayer and reflection are tested and tasted by all we encounter. The contemplative response requires our blood, sweat, and tears. It asks us to die to ourselves again and again and again so that all creation might live. Anything less is useless sympathy. A real contemplative, rather than being passive and unresponsive, is actively and passionately engaged, moved by the Spirit and motivated by love.

Radical Amazement is above all an invitation to live contemplatively, to get caught up in an awe-filled vision that evokes an immediate, active, and compassionate response. We haven't much time. All creation beckons us to respond to the urgent needs that surround us—and they are critical and many. What is required is a radical change in the way we live. Consider these words from the gospel of Luke: "If any one comes to me without hating his father and mother, wife and children, brothers and sisters, and even his own life, he cannot be my disciple" (14:26). These words seem harsh, but I think Jesus, grounded in the reality of first-century Palestine, felt a great sense of urgency about his message and the need to live its truth without compromise. Today, as well, we are asked to leave behind the old paradigm that we learned growing up in Western culture. Jesus' statement is not a judgment about parental methods or family functions and dysfunctions, nor is it an excuse for not taking our relationship commitments seriously. Rather it is a challenge to recognize the signs of the times and to move to a new way of living. We must learn to hate our myopic view of reality. We must turn from the illusions that have given us comfort. We must stretch our imaginations in a way that allows us to see that we are connected to a much larger family than we have ever imagined possible.

Our attitudes and behaviors are rooted in a way of thinking that is no longer reflective of the real. So much of the time we are stuck in the dualistic, hierarchical, either-or thinking that has created the very problems that threaten us. We are not mechanisms with separate parts, but interconnected holons that are mutually dependent. Yet far too often we cling to the individualism and dysfunctional systems that have "parented" us, molding obedient offspring carrying on the "family" tradition in a way that continues to devastate all life, others' as well as our own. Shifting to a new paradigm takes commitment and hard work. It requires gut-wrenching honesty and the willingness to give up fear-filled control. We all know what a difficult undertaking this is, but we are capable of the challenge and perhaps more ready than we think.

A new world view will challenge our old ways of imaging God. A little God—one concerned only for human beings who are like us or should be like us—will never do. The new paradigm includes all creation and is large enough to contain the immensities of the universe. In this new vision we are all connected, all part of a cosmogenesis that continues to unfold, even as you sit here reading these words.

This new world view acknowledges evolution as a creative process urged on from within by the very Spirit of God. It recognizes the special significance of the human species as the consciousness of the cosmos, the universe having emerged in such a way that it is conscious of itself. Along with our consciousness comes the added responsibility to care for Earth and creationkind. Knowing that we cannot possibly contribute to ongoing creation alone, we are empowered by the Spirit, mentored by Christ, and lavished with God's grace.

I have found that the discoveries of science and the new telling of the universe story provide fitting images and metaphors for constructing the new paradigm. They expand my imagination and challenge me to think more creatively. They provide pictures that lure me out of my anthropocentrism and awaken me to a way of living that embraces the entire universe. Science does not seek to "prove" that there is or is not a God—that falls into the realm of theology. But those of us who do believe are not looking for proof, only for images that more accurately reflect reality so that we can live in greater fidelity to what we already know to be true. At the same time, science does not contradict what our faith teaches. While the discoveries may be new, the truths that we will reflect upon here are timeless, found in the depths of Christian tradition and its spiritual practice. As you read I hope you will experience a familiar resonance with the truth.

I think it is important to say some things about my image of God. I use many names for God in these reflections—Creator, Holy One, Mystery, Love, Light—but perhaps my favorite is "incomprehensible Holy Mystery," a term used by Karl Rahner to describe the One who is beyond our images and metaphors, beyond any precise conceptualization. How we image God makes all the difference, determining how we relate to the Holy, who we think we are in ourselves and in the world, and how we negotiate our relationships near and far. In *Radical*

Amazement the image of God that prevails is that of Mystery, of a Holy One who cannot be pinned down to a concrete image but who nevertheless is concretely involved in every piece of life. Saying that God is "within all that is" does not mean that God *is* all (pantheism), but that God is *in* all that is (panentheism), urging all creation to become. God is both transcendent (not to be identified with any particular thing) and immanent (present in all that is). In these reflections, then, I am not equating the universe with God. But the universe also, as God's primary revelation, cannot be separate from God.

Perhaps it is relevant to say something about myself. I am not a scientist. My degrees are in education and religious studies, with special emphasis on Christian spirituality. I am a married, middle class Caucasian woman who is concerned about the Church and its place in the world, and the world and its place in the Church. I am a mother who has concerns for the future of her children and for generations to come. I am a woman who believes in the necessity of prayer and the necessity of action, each informing the other, each essential for the other. This is what I bring to my reflection and writing, just as you bring who you are to your reading. What I desire is that by sharing my own insights you will be supported in your own response to what is significant and challenging in your life. In a very real sense this book is not complete until you have read it and sensed a mutual sharing of hearts that desire the same thing—to live contemplatively and to recognize the truth that the Holy One continues to get us to believe: We all are one.

Radical Amazement contains a series of reflections that build on some discoveries of modern science. Although much of the content of this book has to do with the new science, my purpose is not simply to help the reader acquire scientific facts. There are many other books, much more detailed and comprehensive than this one, which provide that kind of data. My hope is that you will receive this information on a deeper level than the cerebral, that you will take the information and its implications into the core of your being, allowing any remnants of the old divisive paradigm to fall away and a new one to emerge— one that sees connections, affirms life, and transforms the way we live. Or, if you are a person who already sees life primarily in terms of connectedness, my hope is that these reflections will nurture your contemplative spirit as you read and pray with them.

The reflections in this book invite each of us to enter a place of radical amazement, to pay attention to what is being revealed, enabling us to live in more contemplative awareness than we normally do. The discussion of each scientific finding is connected to some awareness from the Christian tradition—whether a scripture or an insight from one of the mystics or other writers—in the attempt to demonstrate that the new science and our religious tradition can experience a resonance that inspires us with wonder. The reflections lead to an invitation to bring the insights into your own life, developing the contemplative awareness that is key not only to personal transformation, but for the transformation of the world.

Each chapter concludes with an invitation to reflection so that the reader may attend to the material more fully and integrate it into life more consciously. *Radical Amazement* seeks to foster the contemplative way, for that is what will save us, that is what will transform us. Each chapter concludes with a prayer that recaptures the discussion and brings us back to the Creator, who is at the heart of all discovery. I encourage you to journal your thoughts and feelings, perhaps having conversations with others who share your desire to emerge from an old worldview and live out of a new paradigm that is lifegiving for all. As more of us share, we create a morphogenic field that encourages and supports others. May we never underestimate our power to change the energy around us, and may we ever remain open to the Holy One who is the source of radical amazement!

CHAPTER ONE

COSMOLOGY

My hand laid the foundation of the earth,
and my right hand spread out the heavens;
when I call to them,
they stand forth together.

* ISAIAH 48:13

How important it is—that we learn the Sacred
Story of our Evolutionary Universe, just as we
have learned our cultural/religious stories. Each
day we will begin to do what humans do best: Be
amazed! Be filled with reverence! Contemplate!
Be entranced by the wonder of the Universe.

* MARY SOUTHARD, CSJ

Knowing who you are is impossible without
knowing where you are.

* PAUL SHEPHERD

From the dawn of consciousness human beings have asked significant questions: "Who are we?" "Where do we come from?" "Where are we going?" Early on, humankind experienced not only a fascination with the celestial bodies that fill the sky by day and by night, but they have known a sense of connection to them as well. Somehow the questions and the stars seem linked, as evident in many of the creation stories that seek to express the origin and meaning of the universe and our place in it.

Every culture in history has told stories in response to the questions that arise out of human awareness and experience of the world. Cosmology is the story that flows out of the study of the origin and development of the universe, including who we are and what we are about. While cosmology includes science, it goes beyond science and the empirical method to explore our experience of purpose and meaning. While cosmology addresses ultimate concerns through story telling by pointing to the Ground of Being, it is not identical with theology, which normally begins with religious experience and engages in a disciplined exploration of human encounter with the divine. Cosmology "is the story of the birth, development, and destiny of the universe, told with the aim of assisting humans in their task of identifying their roles within the great drama."[1]

Stories have power. Because they include metaphor and symbol, the narratives we tell speak to the depth dimension of our lives, shaping our psyches and forming our moral and ethical fiber. Sacred stories—also called myths—express the wisdom of their particular cultures and continue to reveal to listeners some of the profundity of the original experience they describe. Jesus grasped this truth, using parables to teach his followers. Whether invited to consider the response of a father awaiting the return of a lost son or the fate of seed scattered on both good soil and hard stone, those who heard Jesus' stories were drawn in. As they listened, previously unheard truths began to emerge and the paradigm in which they had been entrenched was exposed as deficient. Those who glimpsed the emergent truths were challenged to live out of a fresh structure of reality that broadened their notions about who they were in the world and how they were to live in it.

Besides challenging his listeners to consider who they were, Jesus urged them to consider who God was. Every story Jesus told revealed something of the Holy, the reality he called "Father," the image that itself is a metaphor for what theologian Karl Rahner called "incomprehensible Holy Mystery"—an ultimate reality far beyond any box we try to put "God" in. Jesus' storytelling—his ability to draw folks into another picture of the real—opened up to his disciples and to us the possibility of the reign of God, a way of living that flows out of a vision characterized by compassion and love, a way of living in the here and now, a way of living that is rich, full, and radically amazing.

Since the seventeenth century, when religion and science went their separate ways, we have assumed that the truths and stories of one are incompatible with the truths and stories of the other. The result has been a kind of schizophrenic existence in which we compartmentalize our experience and hope that there will be no clashes that break our carefully constructed categories or expose our illusions. The scripture scholar Anthony Padovano has said that it is not healthy to live and work in one world and believe and pray in another. And yet we do just that. We live and work in a world in which science speaks one version of the truth while we believe and pray in a world in which religion speaks another. And often the two do not seem to relate at all.

If we live and work in a world in which belief and prayer make no sense, if our belief and prayer do not resonate with what we know to be true "out there" in the "real" world, we have nothing to support and nurture us. If we believe and pray in a way that has no contact or

RADICAL AMAZEMENT

connection with the world in which we live and work, our lives can feel meaningless and without purpose. To live and work in one world and believe and pray in another makes our lives seem fragmented and disconnected, even alienated from what is truly lifegiving. I believe the new cosmology can help reconcile the rift between science and religion, enabling us to live more holistically, without the tension and conflict that tear unnecessarily at our lives.

The Old Cosmology

Before looking at the new cosmology in the next chapter, it is helpful to look at its underpinnings. What we know today has been the result of centuries of questioning and searching for answers, an evolution that has proceeded step by step, with each new answer giving rise to the next new question. Here is a brief overview of the foundations upon which the new cosmology rests.

The primary authority on the cosmos in the classical Greek era was Aristotle[2] (384–322 B.C.E.). Because his method was based on actual observation and rational thought, Aristotle is often considered to be the first scientist.[3] In his cosmology, Heaven and Earth were completely separate realms. The heavenly realm embodied the perfect, while the earthly realm embodied the imperfect. In other words, the heavenly and earthly were considered to be two distinctly different kinds of realities. Like the other ancients, Aristotle believed that the Earth was the center of the cosmos, with all the celestial bodies orbiting around her. The Earth was not in motion, but remained fixed and at rest. In this cosmology the orbital paths of the Sun, moon, planets, and stars formed perfect concentric circles around Earth.

Eventually, as techniques became more sophisticated, observers determined that the motion of planets in the sky did not conform to predictions of the circular model. Ptolemy of Alexandria (127–151 C.E.), another Greek astronomer, developed a system of circles within circles that more accurately described the motion of planets. This

Ptolemaic system remained the definitive model of the cosmos for fifteen hundred years.

The first major figure to emerge in the Middle Ages was the Polish astronomer Nicholas Copernicus (1473–1543). Just before his death Copernicus published *De revolutionibus orbium coelestium* (*On the Revolution of the Celestial Spheres*) in which he proposed that Earth rotates on its own axis once each day and revolves around the Sun once each year. Copernicus upset two prevailing views: that Earth was fixed in place and that she was the center of the cosmos. Copernicus still believed that the distant stars were a fixed band of celestial bodies that formed a belt around our solar system.

Although *De revolutionibus* was dedicated to Pope Paul III, it was placed on the Index of Forbidden Books in 1616 and not removed until 1835. Why were the astronomer's ideas so dangerously heretical? Copernicus' revelation was not merely a fascinating astronomical discovery. It changed the way human beings viewed themselves. Our Christian creation stories had been interpreted in a way that placed human beings at the center of the universe, appointed by the divine to have dominion over the created world. What a shock it was to discover that we are not the axis of the universe after all!

It took until the early seventeenth century for Copernicus' claims to be substantiated. In 1609 the Italian scientist Galileo Galilei (1564–1642) fashioned an astronomical telescope and turned it toward the night skies. What he discovered was four moons orbiting Jupiter in the same way that Copernicus had said the Earth revolves around the Sun. Besides confirming Copernicus' theory, Galileo also observed that the planet Venus has phases just like our moon and that the Sun has spots or blemishes on her surface. Through Galileo's observations the notion that the heavenly realm and the earthly realm were two distinctly different kinds of reality began to crumble. The Sun was no longer perfect, and perfect circles no longer accurately described the cosmological system.

As a result of Galileo's work, the distance between Heaven and Earth began to shrink, a development that would have a profound impact on how human beings viewed themselves in relation to the cosmos. By the end of the seventeenth century human beings could no longer regard themselves as the center of the universe, and the heavenly and earthly realms were not so distinctly different after all.

A final discovery made by Galileo was a stepping stone for Isaac Newton. Galileo theorized that the natural state of matter is to move and keep moving rather than remain motionless or fixed. Neither, he proposed, does matter move randomly, but in a straight line at a constant speed. This theory would be crucial to Newton's formulation of the laws of physics.

By coincidence Isaac Newton was born in 1642, the same year Galileo died. It was Newton (1642–1727), an English scientist and mathematician, who discovered the laws of motion and universal gravitation and brought science into the modern era. In 1687 he published *Philosopiae naturalis principia mathematica* (*Mathematical Principles of Natural Philosophy*) or simply the *Principia*, considered by many to be the greatest scientific book ever written. If there were any vestiges of the belief in a perfect Heaven distinct and completely "other than" Earth, they were to crumble under the weight of Newton's discoveries, which revealed that the motion of Earth and all the other celestial bodies operate out of the same laws. The moon and the apple rumored to have fallen on Newton's head are both governed by a force called gravity.

The age ushered in by Newton was one of great scientific discovery and technological advancement. The instruments used for exploration on both the macro and micro levels became ever more sophisticated. The microscope led to the discovery of the atom—which means "indivisible"—leading scientists to believe they had detected the most fundamental building block of the universe. The scientific method that took hold and was used to explore the world during this period involved observation that endeavored to separate things out, reducing them to their smallest basic components, then testing hypotheses through experimentation. Scientific exploration was based on the belief that the universe on both the macro and micro levels consists of discrete units of matter that can easily be disconnected one from another without consequence.

Newtonian physics held that the scientific method allowed scientists to make observations with total objectivity. Scientists thought it was possible to separate the observer from the observed, being completely detached without influencing the observation. The result was that the cosmos was viewed as a machine consisting of perceptible, determined, predictable pieces of matter. This included nature, which became fair game for exploitation in the name of progress, a movement that has had detrimental consequences in our time.

With Newton's discoveries it is possible to see just how deeply the scientific view of reality penetrates our psyches. Just as Newton asserted that the physical arena was governed by certain immutable laws, so Sigmund Freud asserted that the psychological realm is governed by immutable laws. Using the theory of atoms as distinct, boundaried units of matter, the philosopher John Locke proposed that the individual, not community, is primary. Because rational thought and the "objective" scientific method were considered to be the only valid way to view the world, intuition, emotional and artistic expression, and experience of the Holy were devalued as inauthentic and useless, or even harmful.

As these ideas began to penetrate and take hold of our psyches, especially in the Western world, human beings began to change the beliefs that had been a guide in earlier times. As the belief that life is composed of separate units of matter seeped in, individualism replaced the primacy of community. With experience of the divine viewed as suspect, the myths and rituals that spoke to the depth dimension of our lives were assigned at best a peripheral role. Nature lost much of its meaning in human life, no longer viewed as revelation of the divine but as an assortment of forces to be mastered and resources to be consumed. It is no surprise that alienation is one of the chronic diseases of our time, since we have all but lost the sense of connectedness to self, the divine, and nature, all vital components of our fundamental reality.

The scientific discoveries of the twentieth and twenty-first centuries are changing the way we tell the story of the universe. Just as Jesus challenged his listeners to view the world in a way that was different from their customary way, so the changes in the telling of the universe story challenge us to rethink our relationships to the Holy and to those around us. The old story of a static, fixed universe with its mechanistic images, while contributing to the advancement of scientific knowledge and providing a foundation for the discoveries being made today, is no longer viable. Looking at matter as the assembly of separate pieces and parts does not fit with the new revelations regarding the nature of matter. Seeing ourselves as individuals over and above community is illusion. Accurate in many ways, the Newtonian mechanistic view is nevertheless incomplete, a mistake in our understanding of nature which causes a mistake in our understanding of God and ourselves.

We must not underestimate the significance of the way in which we tell the story of the cosmos. Human life is rooted in the practice of sharing the rich variety of our lives—telling stories about who we are and where we have come from, stories that reveal our highest hopes and deepest fears, stories that connect us to one another and cause us to care. It is in the sharing of stories that connections are strengthened, and, conversely, the inability to disclose who we are is a sure sign of a relationship that has broken. One way to look at the discoveries now being brought to light by science is that our relationship with the cosmos has entered a deeper level of self-disclosure. Not only do we know more facts about the universe, we can begin to recognize its movements and intuit the depth of its desires. We have encountered its immeasurable zest for life and its unlimited longing to create. The universe has revealed some of its darkest secrets and most ineffable beauty. When we listen to her story, we cannot help but fall in love—a love that demands a response, a love that asks us to remain connected in a way that is mutually sustaining, mutually caring.

Sometimes, in response to hearing the new story, I disclose to the universe my own. I tell her of my childhood, of lying on my back in the grass at night, looking up at the stars, imagining I could gather them in my hands and toss them all around me like glitter. I tell her that the stars have always made me feel accepted and safe and part of some immense Mystery. I tell her how connected I have always felt to the stars, how one recent morning I awakened before dawn, wrapped myself in a blanket, and watched Jupiter until the daylight caused it to fade from my vision. As I receive her story and she receives mine, we become close friends.

The old cosmology is quickly fading away, but, as Thomas Berry points out, we are still between stories. The old no longer resonates with what we know to be true. The new can make us uncomfortable, even when we are in the midst of radical amazement. Our work, not only for ourselves, but for generations to come, is to integrate the new universe story, to take it in and live it out in all its implications, to intertwine it with our own. I once heard a story from the Native American tradition of the belief that a human being could not reach maturity without making room within herself for the immensities of the universe. That is what watching the stars helps us to do. The universe reveals itself to us, and in the hearing we are transformed. Radical amazement begins with offering hospitality to the cosmos, hearing her story and intertwining it with our own, and nurturing the most significant relationship we have, our relationship with life itself.

WE COME TO CONTEMPLATION

* *What is your story? Imagine that you see your entire life spread out before you. What do you see? Do you notice the old mechanical model of the universe operating anywhere? Do you want things fixed, in place, unchanging? Are there places in your life that you consider separate and apart, not connected to others and to creation? How do feel about that? Is there anything you want to change? Consider how that might happen.*

* *What in this chapter do you find challenging? What do you find radically amazing?*

* *Sit quietly for a moment. What does the Spirit want you to see? How do you respond?*

WE PRAY

Creator God, sometimes we are so accustomed to our habitual way of seeing that we fail to notice what is really there. Like the astronomers of old, we prefer a world fixed in place, with no movement to disturb our treasured vision of reality. Show us the places where we are content to remain set in our ways; show us where we need to be open and changed by new discoveries and new revelations. Help us to expand our awareness and deepen our love for the vast universe that is the ongoing work of your hands. Amen.

CHAPTER TWO

THE NEW COSMOLOGY

The heavens are telling the glory of God;
and the firmament proclaims God's handiwork.
Day to day pours forth speech,
and night to night declares knowledge.
There is no speech, nor are there words;
their voice is not heard;
yet their voice goes out through all the earth,
and their words to the end of the world.

∗ PSALM 19:1–4

We are living through a major period of change in
science, a paradigm shift, from the idea of nature
as inanimate and mechanical to a new
understanding of nature as organic and alive.

∗ RUPERT SHELDRAKE

The heavens are telling the story of God, and all creation stands in awe of the sheer beauty and magnificence of the cosmos. In this chapter, as we review the findings that have brought about the new cosmology, we can also understand these discoveries as the universe telling us her story—and filled with the glory of God that is integral to the telling. Continuously, for billions of years, the universe has told the tale, but only now are we capable of hearing and understanding. Only now can we recognize the speech and hear the voice that goes out through all the Earth and tells us just how radically amazing the gift of life is.

Copernicus, Galileo, Newton, and those who followed them opened the door to the new cosmology.[1] Using ever-more sophisticated instruments, scientists began to discover the vast immensity of the universe. In the nineteenth century astronomers were able to calculate the distance to the moon (about 239,000 miles) and to the Sun (about 93 million miles). In 1838 Friedrich Bessel gauged the distance to a star in the constellation Cygnus to be 6.2 light years (38 trillion miles) from Earth. The universe experienced by humankind was getting bigger!

In 1900 the Milky Way galaxy was considered to be the whole of the universe, with Earth located somewhere within its vast space, although no one knew our precise location. But the universe was about to get larger still. With improved telescopes, astronomers were able to see nebulous clouds of gas within the Milky Way which they dubbed "nebulae." That year, using the new 60-inch reflecting telescope on Mt. Wilson in southern California, scientists were able to calculate that the Large Magellanic Cloud—still within the Milky Way—was 100,000 light years away. Since light travels at the speed of 186,000 miles per second, this is an incredible distance indeed!

In 1924 Edwin Hubble (1889–1953), using an even more powerful 100-inch telescope, confirmed that the Andromeda nebula was really a separate galaxy, not part

29

of the Milky Way at all, containing hundreds of millions of stars of its own. Further, Hubble was able to substantiate the existence of millions of other galaxies. Now we know there are billions of galaxies each with billions of stars. In just a few decades the size of the universe became utterly incomprehensible. And yet more astonishing news was to come.

In the mid-1920s astronomers who were studying the properties of light in space noticed, as they looked at stars from a newly-discovered galaxy, that the light was shifting to the red end of the spectrum of light, that is, to a lower frequency. This phenomenon, called *redshift*, indicated that the stars in the galaxy were moving away from Earth. Soon these scientists noticed that other galaxies were moving away as well—and the ones farthest away were moving at a 60,000 miles-per-second clip! In 1929 Edwin Hubble made an amazing announcement: The universe itself was expanding!

Georges Lemaître (1894–1966), a Belgian physicist and priest, postulated that the expansion implied that there was a beginning to the universe, consistent with the story in Genesis. In the late 1930s Russian-American physicist George Gamow worked out a theory, now known as the Big Bang, which stated that the universe originated in a fiery cosmic explosion from a dense particle smaller than an atom. Judeo-Christian tradition has always held that there was a beginning to the cosmos, a starting point from which all life has originated, and science now confirms that belief.

The Big Bang theory was further substantiated in 1964 with the discovery of microwave background radiation, an energy consistent with an explosive origin. Microwave background radiation enabled astronomers to "see" the universe approximately 380,000 years after the Big Bang. By the 1960s the Big Bang theory had gained wide acceptance among the scientific community as the explanation which best described the origin of the universe based on the mathematical and experimental data available. Although major revisions in the theory have taken place along the way, the Big Bang remains the most widely accepted scenario to date.

In the late 1950s Geoffrey Burbridge was one of many scientists studying radio signals from outer space. What Burbridge discovered was that some of these signals carried the energy of millions of stars. Between 1958 and 1960 Maarten Schmidt, a Dutch physicist, named these energy sources *quasars*—an acronym for "quasi-stellar radiators." Schmidt said that quasars are quite compact, carry tremendous amounts of energy, and are located at the center of galaxies. We now associate these quasars with *black holes*, and astronomers say that every galaxy has a black hole, including our own.

The 1980s saw growth in the understanding of the Big Bang and other phenomena. One of the findings of this era was that galaxies rotate so fast that they should spin apart rather than hold together. This led to the discovery of dark matter, a force which cannot be seen (it neither absorbs nor emits light) but which exerts a gravitational pull that holds clusters of galaxies in place. Scientists say that about twenty-five percent of the universe consists of dark matter.

In 1998 two independent teams of astronomers investigating supernovae (the death explosions of stars), hoping to measure the rate at which the expansion of the universe is slowing down, surprisingly announced that the rate of expansion of the universe is instead accelerating. They called the force behind the expansion dark energy, which, rather than exerting a strong pull like that of gravity, is gravitationally repulsive. Although dark energy has never been seen, support for its existence continues to build:

> Until recently the supernova data were the only direct evidence for the cosmic acceleration, and the only compelling reason to accept dark energy. Precision measurements of the cosmic microwave background (CMB), including data from the Wilkinson Microwave Anisotropy Probe (WMAP), have recently provided circumstantial evidence for dark energy.[2]

About seventy percent of the universe is dark energy. This means that ninety-five percent of the universe is composed of matter or energy that is not visible! Only about five percent of the cosmos is what we call "ordinary" matter.

Finally, in 1998–99, scientists led by Wendy Freedman used the Hubble Space Telescope to collect data to support their research on the accelerating expansion of the universe. This team was able to determine with a fair amount of accuracy the age of the universe. The Big Bang, according to their calculations, occurred approximately 13.7 billion years ago.

While the major scientific findings discussed here are from the realm of astronomy and physics, another significant development that contributes to the current telling of the universe story comes from the arena of geology. In the 1960s James Lovelock and Lynn Margulis postulated that Earth is a self-regulating ecosystem, a biological organism in its own right. Lovelock named this concept the Gaia hypothesis, after the Greek goddess of Earth. He and Margulis pointed to the fact that Earth has maintained a steady surface temperature over hundreds of millions years, just as any biological organism does. Also, salt in Earth's seas remains below the saturation point so that the oceans do not become lethal to plant and animal life, in spite of the fact that each year tons of salt run off into the seas—more evidence that points to our planet as a living, self-regulating system. Rain forests act as lungs and rivers as arteries. Earth, they maintain, is alive, an organism with systems that support her life and ours. Although initially met with skepticism, growing evidence supports Lovelock and Margulis' claim. Steven Goldman explains, "By the end of the 20th century, every aspect of the Earth, from its solid core to the uppermost reaches of its atmosphere, was viewed as 'alive,' continuously driven by the play of awesome forces."[3]

What sweeping changes have taken place during one century of science—amazing discoveries that expanded the magnitude of our cosmos, discoveries that displaced the notion that human beings are at the center of a fixed universe. As these new findings make their way into our lives and psyches, the way in which we think about who we are and who God is will naturally change. Neither human beings nor God are the machines of a mechanistic worldview. Rather, we are holons (to be discussed later in more detail) whose existence is characterized by relationship and connectedness.

The new cosmology offers a way to reconcile the schism that has existed between science and the spirit, between our world of work and our world of prayer. Science and religion no longer have to be at odds in a way that makes us feel that we have to choose one over the other. The new cosmology—the story of the origins and development of the universe and our place in it—will unfold throughout the pages of this book. Like the listeners who heard the parables told by Jesus, we are invited to hear previously unheard truths, brand new pieces of information that may shatter our customary way of thinking and expose our current paradigm as deficient. These new discoveries will challenge us to think about how we live our lives. We will recognize discrepancies and disconnections, we will be invited into the experience of conversion, the opportunity to radically amend how we think and behave, and to set ourselves in a new direction.

Here is a summary of the significant features of the new universe story that have emerged as a result of the recent discoveries of science and which will be the focus of our reflection in this book:

* All creation has come about through a single cosmic event, often called the Big Bang. Creation is not a static, fixed event, but a cosmogenesis, an ongoing act of creation and creativity. Because all life is part of this single cosmic event, all life is connected at its most basic level.

* Evolution as a process that moves toward ever-increasing complexity, and the movement toward consciousness provides a plausible explanation for the development of the universe and its components.

* On a fundamental level energy and matter are interchangeable: $E=mc^2$.

* The theory of holons suggests that everything is a whole/part, that nothing is separate and distinct. Life consists of nested holons of increasing complexity. Relationship is fundamental.

* Surrounding all self-organizing systems there is a force called the morphogenic field. This force somehow organizes information and sets up patterns of thought and activity.

* At the center of every galaxy, including our own, there is a black hole, a region of space governed by a singularity of unimaginable density from which not even light escapes.

* Our solar system came about as the result of a supernova explosion, the death eruption of a primal star. Death is integral to life.

* Seventy percent of the universe is dark energy. Twenty-five percent is dark matter. Only five percent is ordinary matter.

* The universe is expanding at an accelerating rate.

* The language of some scientists engaged in the new cosmology often sounds like the language of the mystics, who acknowledge that our lives are rooted in mystery—and on the level of mystery we are all one.

When I initially encountered various findings of the new science, I was first amazed. Then I began to consider how these breakthroughs would impact my own view of the world, and in particular, my own spirituality, should I seriously integrate them. Would the new contradict the comfortable? Would I be required to change how I think about the world and my place in it? What I found was that the new discoveries resonated with my own experience of the Holy One. More and more my spirituality has been rooted in the belief that unity is God's vision and desire. This has challenged me to uproot many of the habitual ways of thinking and acting that caused me to see myself as separate and disconnected from others. I continue to grapple with the personal implications of the new cosmology, continue to look for vestiges of what Einstein called the delusion of separation, continue to wrestle with my own ineptitude at living the vision. Because the old mechanical view of the cosmos no longer works for me, I pray for the grace to live out of the unity and connectedness that I intuitively know is the truth.

One of my favorite stories about Albert Einstein (1879–1955) involves what he later called the greatest blunder of his life. Steeped in Newtonian physics, Einstein operated out of the world view that the cosmos was fixed, much like a machine. But while doing the mathematical computations that led him to propose the Special Theory of Relativity, he began to see the sweeping implications of his work. If his calculations were correct—and we now know they are—the universe, rather than being fixed, was expanding in all directions. Implied in this insight was the idea that the expansion was away from a single point from which all matter emanates. Einstein, stunned by the implications of his work and reluctant to offer information that would so radically alter what for centuries had been held as truth, fudged his equations! He changed the numbers in order to maintain a static, fixed universe. It took another mathematician, Russian Alexander Friedmann, to call Einstein on his "mistake." Later, in 1931, Edwin Hubble invited Einstein to view the cosmos from his observatory on Mount Wilson, enabling him to see with his own eyes that the implication of his theory was true, that the universe was indeed expanding.

The reason I like this story so much is that even Einstein, the most brilliant scientist of the twentieth century, a great seeker of truth himself, struggled when it came to appropriating a new world view. Knowing the implications—that this new discovery would shatter so much of the "truth" that formed the foundation of his world view—Einstein dug in his heels. How very human he seems, how very much like so many of us who are also reluctant to change the root images, symbols, and beliefs that tell us who we are and where we are going.

Perhaps most of us can resonate with Jeffrey G. Sobosan's words:

How True!

> Some people are intimidated by new knowledge. We get comfortable with the knowledge we have, warm to the values and visions it sustains in our lives and threatened by any proposed refinement, let alone denial of them. The new knowledge comes to us like an assault on cherished idols. [4]

Even Einstein was intimidated by new knowledge. Even Einstein resisted the assault on his cherished world view, making the effort, at least for a little while, to stay in his comfort zone rather than expand the vision and values that sustained him. But eventually he owned that his greatest resistance was his greatest mistake, and from that I take heart.

The new cosmology can upset our old truths as it challenges us to adopt a novel vision of life. Taking a look at a new paradigm will always expose our illusions and bring about a confrontation with our fears. I am reminded of Jesus' impatience with those who fought against the new vision, who had eyes but would not see. And what usually prevents human beings from seeing? Fear. Fear that moves us to grasp for the secure rather than reach for the real. Like Einstein, we can choose to fudge our own equations, living in one world while praying in another. Or we can endeavor to reconcile science and faith within ourselves, allowing them not only a peaceful coexistence but a mutual resonance that permits us to live a life filled with radical amazement.

Brian Swimme says, "Cosmology, when it is alive and healthy in a culture, evokes in the human a deep zest for life, a zest that is satisfying and revivifying, for it provides the psychic energy necessary to begin each day with joy."[5]

The new cosmology—the story that tells how our universe has developed and reveals our unique place in it—is truly and radically amazing. Rather than being contradictory, it affirms and resonates with the heart of Christian spirituality. The Apostle Paul tells us "As it is, there are many parts, yet one body." (1 Cor 12:20) The universe consists of many parts, yet all that is remains connected at the most fundamental levels. Living with sensitivity to the connectedness at the root of our being makes everyone family, progeny of the Creator who shapes and fashions and breathes life into us all. This new cosmology, alive and well, can serve to unite us all, for it is the story of us all. No one is excluded; everyone has an essential role in this great tale of Love.

WE COME TO CONTEMPLATION

* *In this exercise, you are invited to engage your imagination. If it is helpful, sit relaxed, with your eyes closed. First imagine Copernicus working over his calculations and coming to realize that Earth is not the center of the universe but actually revolves around the Sun. . . . Consider his astonishment, his fear, his uncertainty. . . . Invite Copernicus to tell you about his experience and listen to what he says about the challenge of changing belief systems.*

* *Next, imagine that you are in conversation with Albert Einstein. Let him tell you about his "blunder," when he altered the equations so that they would not reveal that the universe is expanding. . . . Ask him about his fears and concerns. . . . How would he have done things differently? What recommendation does he have for you?*

* *Now bring your own thoughts and feelings into conversation with Jesus or other wisdom figure in your life. Are there places within you or circumstances in your life that invite you to*

change? What goes on within as you consider something as significant as a paradigm shift—not just changing an idea or notion, but really altering the root beliefs that you have lived out of for your entire life? Where is the resistance—and how do you want to respond to it? What do you have to lose? What will you receive if you say yes? What response do you hear from Jesus or another wisdom figure that is helpful for you?

* What in this chapter do you find challenging? What do you find radically amazing? *not like close-minded Interconnectedness — actions*

* Sit quietly for a moment. What does the Spirit want you to see? How do you respond? *So many things that we worry about are not worth worrying about.*

WE PRAY

Holy God, Creator of the Universe, we know that you are Mystery beyond mystery, yet our finite minds and fragile fears often want to contain you in a way that keeps us comfortable. We resist stretching our beliefs to include the new. We hesitate to break with the past, even if it is filled with illusion. Help us to be open to your revelation, not merely because it is new, but because it can bring us closer to the truth of who you are and who we are in you. As we encounter your mystery and love in the hearing of the new story of creation, may we be open to radical amazement and led into a way of living that embraces the connectedness of all that is. Amen.

CHAPTER THREE

IN THE BEGINNING: THE UNIVERSE FLARES FORTH

In the beginning when God created the heavens and the earth, the earth was formless and void and darkness covered the face of the deep.

✳ GENESIS 1:1–2

The fire which is in the sun, the fire which is in the earth, that fire is in my own heart.

✳ MAITRE UPANISHAD 6:17

In the beginning. The phrase is so simple that we can brush past it with little notice. But that primordial beginning is not merely an event that makes for interesting science. It marks our birth, too. Everything we are, have ever been, or hope to be is rooted in the moment when time and space began and life burst forth. Imagine being able to film the universe as it expands—then watching the rewind to see all of the separate parts converge at a single point. This is what the Big Bang theory suggests, and it is a fundamental part of the new cosmology.

Here is a brief summary of the story of the cosmos as it is now being told: The universe came into being in a flaming cosmic explosion that gave birth to space and time. By measuring such phenomenon as cosmic microwave background radiation, scientists have determined that the birth of the universe as we know it occurred about 13.7 billion years ago. All the energy that has ever existed—and therefore all the matter that has ever existed—was formed in this single spectacular cosmic event.

Ours was a fiery birth, an unfurling of vast, raw potentiality which quickly began to take shape and form. The dense fireball that erupted in less than a trillionth of a trillionth of a second was a billion trillion times hotter than our sun. In the first millionth of a second, the fundamental principles of the entire physical universe emerged, including the laws of gravity and electromagnetism. By the first second the universe consisted of fundamental particles and energy. The collision of photons formed protons, neutrons, and electrons, the basic components of atoms. Energy was already converting to mass. By the time three minutes had passed, when the universe had cooled to one billion degrees, protons and neutrons had come together to form hydrogen and helium, elements which would eventually fuel the first generation of stars.

The timing of all this birthing is nothing short of miraculous. If the unfolding had been a trillionth of a trillionth of a trillionth of one percent faster, the cosmic material would have been flung too far apart for anything significant to happen. If, however, the unfolding had been a trillionth of a trillionth of a trillionth of one percent slower, the universe would have collapsed in upon itself.

Ten thousand years after the Big Bang most of the energy in the universe was in the form of radiation. For the next three hundred thousand years the universe continued to expand and cool until the amount of energy in the form of radiation and the amount of energy in the form of matter were about equal. This made possible the next major development. Three hundred million years after the Big Bang the first stars and galaxies emerged, forming clusters as the universe continued to expand.

The next phase of the unfolding of the universe was extraordinarily turbulent. Great galactic collisions took place, setting off cosmic explosions of gas. Some galaxies survived, while others did not. Some continued to create stars, while others did not. It was in these first stars that the primary elements were formed: carbon, oxygen, nitrogen, copper, silver, silicon, magnesium, calcium, sulfur, and iron—all the elements that are components of life today.

The flaring forth continued. Stars aged, then died in great supernova explosions. A supernova is the death eruption of a star, which happens in a dazzling cloud of light, followed by a gradual fading. We know from seeing images from the Hubble Space Telescope just how brilliant these supernovas are—radiating more light than an entire galaxy of stars. The birth of our own solar system—Sun, Earth, our neighbor planets and planetoids—came as the result of a supernova explosion that occurred about five billion years ago.[1]

What a beginning! What a miracle that we are here, both as a species and as persons. Since all the matter that now exists was formed in that initial bursting forth of life, we must each be formed out of the primordial dust that

42

became stars. Our true moment of birth cannot be measured in decades. We are 13.7 billion years old! We carry within us the very energy that fashioned the stars. We carry within us the evolution of the elements, eons of development that resulted in who we are today. We carry within us the fire that flared forth into life. As Bede Griffiths put it,

> The explosion of matter in the universe fifteen billion years ago is present to all of us. Each one of us is part of the effect of that one original explosion such that, in our unconscious, we are linked up with the very beginning of the universe and with the matter of the universe from the earliest stages of its formation. In that sense the universe is within us.[2]

How accustomed we are to thinking of the universe as something "out there," something accessible only by telescope or rocket ship, something completely separate from us. And now we know that this is not the case. Energy in the form of matter has been developing, re-arranging, reconstituting, evolving, growing, transcending since the beginning of spacetime itself. Think of this: In its latent potential, the embodied person that you are at this very moment—all the constituents that would eventually come together into the person that is you—was present in the Big Bang. Radically amazing!

The new narrative of the unfolding of the universe does not contradict our Christian story, but actually resonates with it quite well. Our faith tradition has always maintained that there was a beginning—a moment in which time began, in which a mighty and holy power hovered, brooded over, and birthed all that is. We believe all creation has come about through the movement of the Mystery we call God. We believe not only that God initiated the creation event, but that divine presence flows in and through the experience of ongoing creation. Although the Creator cannot be reduced to the creation event or the universe that has been formed, neither is God separate from all that is. The Creator's presence in the form of life itself is woven into the fabric of the universe, apparent to all who take the time to see.

One of the truly amazing discoveries of all time comes not from science, but from the experience of simple people from simple places who have been able to see, who have been able to recognize God's presence in the simplicity of their lives. They have recognized that the Holy One says not only to Jeremiah, but to each of us, "Before I formed you in the womb I knew you" (Jer 1:5). We are known by the God who, in some mysterious and magnificent way, willed life into being—and with it our particular being. With those who have gone before us in faith we can pray

> Where can I go from your spirit?
> Or where can I flee from your presence?
> If I ascend to heaven, you are there;
> If I make my bed in Sheol, you are there.
> If I take the wings of the morning
> and settle at the farthest limits of the sea,
> even there your hand shall lead me,
> and your right hand shall hold me fast.
> If I say "Surely the darkness shall cover me,
> and the light around me become the night,"
> even the darkness is not dark to you;
> the night is as bright as the day,
> for darkness is as light to you. (Ps 139:7–12)

Not even darkness can hinder the God who is light, whose presence permeates every moment of our lives, no matter whether we are in the heights or depths of human experience. In a way that remains a mystery, the Holy One overcame the darkness of nonbeing, bursting forth as light and life that is unceasing in its dynamism and creativity. *That* is the reality that underpins the life of the universe and each creature's life as well.

In the *Spiritual Exercises*, Ignatius of Loyola invites the one who prays to "Consider that God, your benefactor, is present in all creatures and in yourself. If you look at every

RADICAL AMAZEMENT

step of the visible creation, in all you will meet God."[3] God, Ignatius goes on to say, is in the elements, giving them existence; in plants, giving them life; in animals, giving them sensation. God, he concludes "is in you; and, collecting all these degrees of being scattered through the rest of . . . creation, God unites them in you."[4] Written in the sixteenth century, these words echo with truth in a vibrant new way when heard by twenty-first century ears. Ignatius taught that it is possible to find God in all things, that all creation pulses with the life that has its origin in God. Science affirms just how integrally connected we really are.

Even as we acknowledge that God is in all things, living in awareness and operating out of attentiveness to that reality is another matter. If we do not cultivate attentive awareness we will miss the divine presence springing forth all around us in unrestrained beauty and power. If we do not have moments of pause, we will be incapable of the long loving looks that make us springs of beauty and power who are forces of compassion and service in the world. Fostering a contemplative practice and perspective are critical components of co-creativity.

A spiritual practice that many find helpful is Christian meditation, or centering prayer. This discipline is rooted in centuries of experience and allows us to support and maintain a contemplative stance. The basic elements are simple: Sit straight and still, allowing your eyes to close or lower to the ground and your hands to rest comfortably on your lap. Breathe slowly and regularly, being aware of your abdomen filling and emptying. Some practitioners find it helpful to use a mantra or sacred word or phrase such as "fill me," "free me," "I love you," or "come, Holy Spirit." These words are not spoken aloud but repeated interiorly and in harmony with breathing. The choice of a word or phrase is not as important as what the repetition of the word or phrase does—slowing the breathing and allowing the mind to grow still so that the body relaxes and incessant thinking diminishes or stops. James Finley recommends that as we sit, we remain "present, open and awake, neither clinging to nor rejecting anything."[5] Thoughts will come—all the things we plan to do or want to do or forgot to do or choose to worry about—but we simply let those thoughts go without judgment and return to our breathing or phrase.

When we first begin a contemplative practice it is easy to become frustrated with all the thoughts that bombard our prayer. Do not become discouraged. "Success" and "doing it right" are not significant here. What is significant is remaining faithful to the process of prayer and allowing your body, mind, and spirit to come into harmony. This normally takes some time. Be assured that even when you sit and find yourself assailed by an unruly mind or fidgety body, you will still experience benefits from your fidelity to prayer. You will find yourself more peaceful and more compassionate with the world around you. If this practice is unfamiliar or difficult for you, consult with a spiritual director or someone trained in facilitating this kind of prayer. James Finley and Thomas Keating each have books and audio recordings that are helpful as well.[6]

As we envision the Big Bang and the birthing process that it ignited, we are awed by the utter incomprehensibility, the utter grandeur of creation—and the Creator who is birthgiver and sustainer. Like the psalmist we cry out "How marvelous are your works, O God!" or "How radically amazing!" Each and every one of us is the result of 13.7 billion years of hovering by the God whose name is Love, the Holy One who continues to shape and fashion, continues to call us to live deeply awed and intensely committed as we participate co-creatively in this radically amazing story. Our life is rooted in and connected to all life—all that has been, all that will be. We are not final products, end result of a single act. We are participants in a single, on-going act of Creation.

WE COME TO CONTEMPLATION

* *Close your eyes and imagine in slow motion the beginning of life—the brilliance of the initial*

RADICAL AMAZEMENT

flaring forth . . . the transparency . . . the formation of galaxies and their collision as stars emerge . . . the formation of elements . . . the supernova explosion that gave birth to the planetary system that is our home . . . the formation of Earth and all her creatures. . . . Imagine the Creator who directs this process, who continues to hover and brood and birth all life.

* *Slowly read the first two verses in Genesis, allowing any new awareness to arise: In the beginning when God created the heavens and the earth, the earth was formless and void and darkness covered the face of the deep.*

* *How do you describe or image God? Does this view of creation change how you think of God? How does it change how you think about yourself and all others?*

* *What in this chapter do you find challenging? What do you find radically amazing?*

* *Sit quietly for a moment. What does the Spirit want you to see? How do you respond?*

WE PRAY

Holy One, you who fashioned time and space itself, we are in awe of your work, radically amazed by your ceaseless creativity and love. We see your hand in the first flaring forth, your revelation and grace in each successive step along the way. Even now you manifest your presence in this moment as it unfolds, here, now, and permeates all creation. Help us to grow in our ability to attend to you and your presence in all that is. May our contemplative awareness deepen and allow us to see more clearly your creativity and your love. Amen.

CHAPTER FOUR

LET THERE
BE LIGHT:
THE BIG BANG

And God said, "Let there be light"; and there was light.

*GENESIS 1:3

God is the living light in every respect.
From God all light shines.

*HILDEGARD OF BINGEN

Before
the Holy One created
there was only darkness:
no thing,
formless void,
fecund emptiness
unexpressed.
Then
there was
Light—
fiery pentecost
from formless dark,
flaring forth,
refulgence
unabashed.
Through this Light
came all that is.
And God said
"Life is *very* good."

Before the great creative event there was nothing—only what we imagine to be darkness—and then the great primordial fire let loose, a fiery furnace in which all that exists was forged. As it turns out, the process described by the early storytellers of our tradition is not at odds with what scientists say really happened. Astronomers believe that 10–20 seconds after the Big Bang—almost instantaneously—the first photons emerged as elementary particles collided. Brian Swimme and Thomas Berry call primal event the great "flaring forth," an image filled with creativity, fecundity, and power. At the moment of birth, the universe was bathed in light.

It is helpful to think of light, a form of electromagnetic radiation, as waves, like ripples on the surface of a pond. The energy of light travels at a fixed speed and is measured in wavelengths, the distance between the peaks of the ripples. The longer the wavelength, the lower the energy it contains; the shorter the wavelength, the higher the energy it contains. The ordinary way we distinguish the energy of light waves is in terms of color. Blue, indigo, and violet are indicative of shorter waves of light and higher energy, while red and orange waves are longer and display lower energy. Green and yellow carry intermediate wavelengths and energy.

Most of the time when we speak of light we are referring to that which is visible, but in actuality most light falls outside the visible spectrum. Light with wave lengths even shorter than violet and blue include ultraviolet light and X-rays. Light at the other end of the spectrum, with very long wavelengths, include infrared rays and microwave radiation.

Significant discoveries in the early twentieth century provide the foundation for our current understanding of light. Perhaps the most interesting property discovered about light is that it possesses both the characteristics of a wave and the characteristics of a particle. In 1900 Max Planck (1858–1947) proposed that electromagnetic energy

49

(light) did not travel in a straight, steady flow as it seemed. Rather, Planck asserted that light travels in discrete packets of energy he called "quanta"—what we now call photons. Light can be measured both as particles and waves.

In 1905 Einstein introduced the Special Theory of Relativity and one of the most famous equations in the history of science, $E=mc^2$. This equation was a radical breakthrough because it signifies that energy and mass are convertible: Energy (E) equals mass (m) multiplied by the speed of light (c) squared. But this is not what Einstein started with, but rather where he ended up.

Einstein based special relativity on two postulates. The first is that the speed of light is constant—186,000 miles per second—regardless of whether or not a body is in motion and regardless of its frame of reference. This seems counter-intuitive. We know that if we are in a vehicle that is moving at fifty miles per hour and another approaches us at fifty miles per hour, we experience the vehicle coming toward us at one hundred miles per hour. If we stop, however, and the other vehicle continues at the same speed, we experience its approach at fifty miles per hour. The speed or velocity of an object is affected by the movement of each vehicle. This does not hold true with regard to light. When measuring the speed of light, it doesn't matter if we are moving toward the light or at rest. The speed of light measured will always be 186,000 miles per second.

The second important postulate, the principle of relativity, states that all the laws of physics are equally valid in all frames of reference moving at a uniform velocity (i.e., frames of reference that are not accelerating or decelerating). From these two postulates Einstein derived a number of strange features of the physical universe, all later demonstrated in experiments. The first, the convertibility of mass and energy, has already been mentioned. Beyond that, Einstein also showed that quantities we assumed to be absolute—time, length, and mass—are not in fact so. Strangely enough, a spacecraft that blasted off for Proxima Centauri and traveled there at ninety-five percent of the speed of light would become significantly shorter to the observer on Earth while maintaining the same length when measured by those traveling in the spacecraft. Meanwhile on Earth 4.5 years would elapse before the ship arrived at its destination 4.35 light years away. But the extremely precise atomic clock on the ship would show just 1.4 years passing. Amazingly enough, the answer to "How long did that trip take?" would depend entirely on who was asked! Meanwhile the mass of the spaceship as it approached the speed of light would increase.

In the 1940s scientists working on the formulation of the Big Bang theory, led by George Gamow, suggested that if the Big Bang theory were true, there should be a sort of cosmic afterglow in the form of microwave radiation uniformly distributed throughout the universe. The ability to detect and calculate microwave background radiation could be a way to confirm their theory and perhaps pinpoint more precisely how long ago the Big Bang happened, if indeed it had occurred. Unfortunately, at the time there was no technology available to detect such a phenomenon.

In the 1960s, however, at the dawn of the telecommunications era, two engineers working at Bell Laboratory, Arno Penzias and Robert Wilson, developed an ultrasensitive antenna that could pick up signals at the microwave range from Telstar, the first communications satellite. There was a problem with the antenna, Penzias and Wilson discovered, because no matter which way they pointed it there was a constant hiss in the background. Yet, try as they may, the scientists could find no source of equipment malfunction to explain the static. Serendipitously, they ran across a paper written by a researcher at Princeton, James Peebles, who had been working on the Big Bang theory and the possibility of cosmic microwave background. Penzias and Wilson soon realized that the hiss their antenna was picking up was actually remnants of the Big Bang! This discovery was so significant, putting the Big Bang theory front and center once again, that in 1978 Penzias and Wilson were awarded the Nobel Prize. Since that time researchers have been able to "see" the universe as it was about 380,000 years after the Big Bang—very early in its 13.7 billion year history. These are the oldest pictures of the universe to date.[1] Ironically, light, the first manifestation of the created universe, is also the tool by which the mystery of the universe is being revealed.

The discoveries about the nature of light made in the past century provide an image for explaining how integral the Holy One is to creation. In the Genesis narrative, the first words spoken by God are "Let there be light." The Creator speaks and it is so. Gerald Schroeder writes:

> It is highly significant that light was the first creation of the universe. Light, existing outside of time and space, is the metaphysical link between the timeless eternity that preceded our universe and the world of time space and matter within which we live.[2]

Light, according to Schroeder, can "abandon the ethereal timeless realm of energy and become matter. In doing so, it enters the domain of time and space."[3] What an appropriate metaphor it is for the Creation event. In the Big Bang, light manifests and begins immediately to organize and take shape, eventually becoming the matter that forms stars and galaxies, Earth and her mountains and streams, human beings and all creationkind. Light, as expression of the divine breath and spirit, links Mystery and matter. Mystery and matter are inextricable.

Because Mystery and matter are entwined each reveals something of the nature of the other. Just as light in the form of microwave radiation pervades ordinary space and allows us to picture the original creation event, so the glimmers of light that permeate our daily experience allow us to catch a glimpse of the Creator behind the originating event. Matter tells us that Mystery is endlessly and ceaselessly creative. The colors around us in all their subtle hues are evidence, each a different wavelength of the same

RADICAL AMAZEMENT

vibrant light. Matter, having come from light, reveals that all life is connected to the one Light, the source of all being. Conversely, Mystery tells us that matter is a significant expression of its very self, and therefore all matter is holy, all matter is embodied Light. Mystery, having poured itself out endlessly and ceaselessly in creation, longs for us to recognize the connectedness and live in fidelity to that truth. We are beings of Light and to live without recognizing this truth is to live in darkness indeed.

It is no wonder that our scriptures are filled with images of light. We acknowledge God as our light and salvation (Ps 27:1), the source of life itself. We recognize that "The unfolding of [God's] words give light" (Ps 119:130) and that our light and life proceed from the light and life of the Holy One. Those who acknowledge the connectedness between Mystery and matter are said to walk in the light (Is 2:5), and when we relate to the divine in a way that recognizes the connectedness of all that is, our light—God's light within us—will break forth like the dawn (Is 58:8). In a crucial way, as we will explore in a subsequent chapter, Jesus is the light who teaches us about how to live as beings of the Light.

One of the difficulties that we may have in the spiritual life comes from knowing that divine Light, like the material manifestation of light, is most often not visible, lying beyond the ability of our unaided eye to detect. Do you remember playing childhood games with a handkerchief tied over your eyes and peeking through little gaps to see what was around you? You knew something more was there, but it was not within your field of vision. You could hear the voices of playmates and other sounds that indicated something was going on. But you couldn't see anything. Do you remember how wonderful it was to pull the blindfold off? Sometimes, if the handkerchief was tight, the light that streamed in was almost blinding.

In the physiological realm we cannot naturally alter our capacity to see beyond a limited range of light, but in the spiritual realm we can. Contemplation, both as a committed practice and an integrated lifestyle, broadens our scope of vision, increasing the spectrum of what is visible to us. The experience amounts to the spiritual equivalent of receiving new eyes and seeing what has been integral to the universe all along. It changes the intensity of the experience of spirit by removing the illusion of distance and affirming that the Spirit is present from within each moment and particle of our existence.

Elizabeth Johnson says, "The act of creation is already a Pentecost, a first and permanent outpouring of the fiery Spirit of life."[4] The new telling of the universe story affirms such a notion. This fiery Spirit, the energetic force that shapes and fashions all that exists, is divine Light, holy Energy. Peter Hodgson writes, "Spirit is an immaterial vitality that enlivens and shapes material nature. . . . Energy is simply that mysterious power that is active and at work in things, and that power is God as Spirit."[5]

In contemplative moments we experience the power of God as Spirit, as Light, as Life, as Creator, as Divine Energy that shapes and fashions us. There is not one fragment of life, not one particle or wave of existence that is separate from this reality. God is the light of our being and of all being, and each wave manifests the power of God's Spirit. How can we respond, other than to stand in awe and wonder of just how integrally the Spirit of Light is part of us? Each of us, without exception, comes from Light. Radically Amazing!

WE COME TO CONTEMPLATION

* *Imagine that you are present at the beginning of time, present when the great flaring forth occurred. In the first split second, imagine that photons erupt into being and are inseparable from the divine energy we call Spirit. . . . See the photons carry light outward as the universe expands, recognizing that the expansion of the universe is really the expansion of Spirit. . . . Watch as stars and galaxies begin to emerge—*

RADICAL AMAZEMENT

still manifestations of Spirit. . . . Envision Earth and all her creatures as manifestations of Spirit. . . . See yourself, your body, as being composed of light—light from the initial Big Bang, light that is inseparable from Spirit. . . . Allow yourself to be amazed at how the Spirit of God is part of every cell of your body, every part of your consciousness, every piece of your experience.

* *Sit in the sunlight for a few moments, gathering the light into yourself, considering the fact that you have come from Light.*

* *How does knowing more about light and how integral it is to life affect your awareness of God? How do you image God differently? What* awesome! *changes? What remains the same? What do you want to affirm about your relationship with God?*

* *What in this chapter do you find challenging? What do you find radically amazing?*

* *Sit quietly for a moment. What does the Spirit want you to see? How do you respond?*

WE PRAY

Divine Light, at no time have you ever been separate from all that is. You do not stand outside of creation, but are the very light and life within us. Just to be is to exist in and through your love. This insight amazes us and helps us to see not only how deeply we are connected to you, but how deeply we are connected to one another. Divest us of any illusion that we are separate from others. Help us to live in awareness of the sacredness of those connections and to express our awareness in loving attitudes and behavior. Thank you for revealing yourself as you do, for using science to bring us to contemplative awareness of who you are and who we are in you. Amen.

CHAPTER FIVE

ALL CREATION IS GROANING: THE PROCESS OF EVOLUTION

For the creation waits with eager longing for the revealing of the sons of God; for the creation was subjected to futility, not of its own will but by the will of the one who subjected it, in hope that the creation itself will be set free from its bondage to decay and will obtain the freedom of the glory of the children of God. We know that the whole creation has been groaning in labor pains until now; and not only the creation, but we ourselves, who have the first fruits of the Spirit, groan inwardly as we wait for adoption, the redemption of our bodies.

✳ ROMANS 8:19–23

Our ancestry stretches back through the life forms and into the stars, back into the beginnings of the primeval fireball. This universe is a single multiform energetic unfolding of matter, mind, intelligence, and life.

✳ BRIAN SWIMME

We are participants in a grand unfolding of life that is eons old. What makes human beings unique but not better than the rest of creation is our capacity for self-reflection, for a quality of consciousness that allows us to know that we know. It appears that only humankind has this capacity. We alone can comprehend the unity that is at the core of the cosmos. We alone can be radically amazed at the work of creation.

Teilhard de Chardin, citing Julian Huxley, said that human beings "are nothing less than evolution become conscious of itself."[1] The consciousness of each of us is the result of the evolution of consciousness which has proceeded for eons. In us the evolving universe is capable of self-reflection. What a profound insight, one that compels us to look at ourselves in a much larger context. We are the universe conscious of itself. A developmental process of 13.7 billion years—unfolding one tiny evolutionary step at a time—enabled us to become *Homo sapiens*, the ones with the ability to be aware.

In *Reclaiming Spirituality*, Diarmuid O'Murchu says that the quality of consciousness human beings possess "contrary to being a special endowment with which we seek to lord it over the rest of creation, needs to be freshly understood as an integral dimension of the 'intelligence' that permeates all life in the universe."[2] What we as human beings envision, what we dream and desire, what we hope for and work toward—all of it affects the universe, all of it has its impact on Earth and every single creature that is. Our gift of consciousness is both amazing and demanding. Our capacity for awareness is both central and essential.

At the heart of the passage from chapter eight of Romans is the awareness of connectedness. All creation, together as a whole, awaits freedom from bondage. All creation groans for wholeness and stretches forward in transformation. The language of this passage from scripture suggests a birthing process, one filled not only with travail but with eager longing for a fuller revelation of God in our midst. All creation is mysteriously bound together in a process of becoming all that we hope for.

In the gospel of Mark, Jesus' final words to his disciples encourages them to "go into all the world and preach the good news *to the whole creation*" (Mark 16:15). The good news of the gospel—the new vision—is intended for all creation, not human beings alone. Why? Because human salvation and freedom cannot be separated from the salvation and freedom of all creation. Our fundamental connectedness will not allow it. Who we are cannot be separated from where we have come from and where we are going. Human beings are the result of billions of years of development that has proceeded one evolutionary step at a time. Who we are flows from all that has gone before, the step-by-step emergence from elementary subatomic particles to primitive unicellular structures to multicellular organisms to hominids to human. We are not so different from other species as we might think. Recent genome research has concluded that of the 30,000 genes in the common house mouse (*Mus musculus*), ninety-nine percent have direct counterparts with the human species.[3] Despite appearances, we all share essential components of life that make us more similar than different.

Thomas Aquinas said that the whole universe together participates in the divine goodness and represents it better than any single being. He recognized that creation is a whole and that all life shares in the experience of divine grace. From modern science the theory of evolution offers an image that resonates with Thomas' insight and enables us to grasp the fundamental principle that all life is a manifestation of divine goodness.

Evolution, David Quammen wrote in *National Geographic*, is "a theory about the origin of adaptation, complexity and diversity among Earth's living creatures."[4] Perhaps it is first helpful to clarify just what is meant by a "theory." Copernicus' idea that Earth orbits around the Sun is a theory. The existence of atoms is a theory. We plug in a lamp and switch on the light bulb based on

RADICAL AMAZEMENT

electromagnetic theory. Drawing on the theory of aerodynamics we build airplaines and fly to destinations all over the world. A theory is "an explanation that has been confirmed to such a degree, by observation and experiment, that knowledgeable experts accept it as fact."[5] A theory is not just a conjured-up idea that has no basis in reality. It is an explanation of how the universe or its components work that is consistent with the evidence and sensible to the specialists in that particular field. Theories themselves evolve and are refined or even eliminated as new data accumulates.

Much has been made of the potential conflict between evolution and faith. Dogmatic materialists wield the theory to try to wrench humankind out of any special relationship with the Divine while fundamentalist Christians hold that the Bible and evolution are incompatible, and so much the worse for the theory of evolution. Should we worry about some grand clash between faith in reason? In 1950 Pope Pius XII, in the encyclical *Humani Generis*, affirmed that there is no conflict between the theory of evolution and the Catholic-Christian faith tradition. In 1996, in a message to the Pontifical Academy of Sciences, Pope John Paul II declared

> Today, more than a half-century after the appearance of [*Humani Generis*], some new findings lead us toward the recognition of evolution as more than just a hypothesis. In fact it is remarkable that this theory has had progressively greater influence on the spirit of researchers, following a series of discoveries in different scholarly disciplines. The convergence in the results of these independent studies— which was neither planned nor sought— constitutes in itself a significant argument in favor of the theory.[6]

According to Pope John Paul II, information gathered by science is not inconsistent with long-held theological truths. In addition to the Roman Catholic tradition, officials of most mainline Protestant denominations agree that the evidence for evolution is compelling, giving followers the latitude to decide what they will believe in this matter. The point is that the evidence for evolution is substantial, and mainstream religious traditions are not opposed to the theory because they do not consider it a contradiction to the Biblical creation narrative. In fact we will see that to take in the theory of evolution—to step back from any fear or resistance—and see the simple beauty of how creation unfolds can bring us to the brink of awe and catapult us into the Mystery that is the source of all.

Although the theory of evolution was not new, it was Charles Darwin (1809–1882) who offered an explanation of how evolution occurs. In his *Origin of Species*, published in 1859, Darwin described the process of natural selection. According to the theory of natural selection, tiny random changes that can be passed on to the next generation occur in all species. These changes may influence components such as size or shape, biochemistry or behavior. As mutations occur, they affect the ability of members of the species to survive and reproduce, and the members that thrive and reproduce most abundantly eventually suppress weaker, less flourishing members. This process, called anagenesis, leads to the transformation of a species over a long period of time. Eventually changes can occur to such an extent that a new species evolves. (A new species is one that is unable to reproduce with the first species.) This process is called speciation.

Darwin drew from biogeography (which species live where), paleontology (extinct life forms), embryology (development prior to birth), and morphology (anatomical shape and design) to substantiate his theory. Today's researchers have even more methods for gathering information. The sciences of population genetics, biochemistry, molecular chemistry, and genomics all lend support to the theory that has developed over the past century and a half. Recent adjustments to the theory affirm

RADICAL AMAZEMENT

the awareness that no species is isolated, that each species develops as the result of dynamic relationships with other species and changing environments. The theory of evolution remains the best explanation of how life has developed as the experts in the various scientific fields gather evidence. Indeed, much of the current research in medicine, notably the study of viruses and the development of drugs that save lives, is based upon the theory of evolution.

Does this theory contradict what our faith tells us? Not at all. Does the notion that we descend from common ancestral organisms diminish the fact that we human beings are the universe come to consciousness of self? No. Rather than diminishing the human species, the theory of evolution can affirm that we are related to all creation, kith and kin to every species that inhabits the face of the Earth, with ancestors that span eons and with a trajectory that can only become more and more amazing. I often wonder if resistance to the theory of evolution is yet another unconscious incidence of the reluctance to surrender our anthropocentrism and an attempt to retain the ancient illusion that the universe is only about us humans. Such resistance in the face of substantial evidence to the contrary binds humans as well as all species.

No wonder that all creation groans together for freedom. No surprise that all creation stretches as one body, always pregnant, always struggling to give birth to the new. All creation participates in the creative Mystery together. The fecund energy we call God is at the heart of the evolutionary process, working from within creation, endowing creatures themselves to be empowered to choose life in whatever way is appropriate for their mode of being. All creation experiences this divine movement from within. Together with all creation we are partners in transcendence and transformation, which makes us, in this sense, partners in salvation. This is not to say that individual or personal salvation is not significant and essential, but the connectedness demonstrated by the theory of evolution makes salvation impossible without including and acknowledging the whole of creation. This tells us that the Body of Christ, rather than simply being a group of like-minded human beings, includes all of life. Paul writes

> As it is, there are many parts, yet one body. The eye cannot say to the hand, "I have no need of you," nor again the head to the feet, "I have no need of you." On the contrary, the parts of the body which seem to be weaker are indispensable (1 Cor 12:20–22).

The theory of evolution and the new universe story take Paul's words to greater depth. Now we know that the human cannot say to the river, "I have no need of you," nor to the creatures who inhabit the forest, "I have no need of you," nor to any particle of creation anywhere in cosmos, "I have no need of you." All is sacred, all is expressive of the Holy, all is a manifestation of the divine. All are joined together in the unity of Love that is creation itself, and all are called to participate in this unity in and as the creation it is, no matter what its complexity or consciousness may be. Even those pieces of creation we consider insignificant are, according to Paul, indispensable to our life and to our salvation. Each member of creationkind brings with it a gift for all, and our refusal to accept the gifts of any diminishes the whole.

Christian tradition has always been clear that any gift we have is not for ourselves alone but for edifying one another in community. The gift of human consciousness and all the capacities that accompany that gift are not only for human beings, but for creation as a whole. That is our role in the universe, that is what gives us purpose and meaning. Salvation, at its root, means to be whole. And since we cannot be whole without acknowledging all of the parts that make us one, our salvation—our own wholeness—is intricately bound to the salvation of all. Our salvation as planetary creatures depends upon acknowledging and receiving the gifts of mountains and their streams, of woods and meadows, of clean air and undamaged soil, of ecosystems that keep the planet and human existence both viable. Salvation that is only about "me" or an exclusive "we" is not salvation at all.

In *Emergence: The Shift from Ego to Essence*, Barbara Marx Hubbard writes that humankind is at a crossroads in the process of evolution. Ours is a unique and urgent time in which we are challenged to make the transition from *Homo sapiens* (wise ones) to *Homo universalis* (universal ones). What is unique about this evolutionary moment, Hubbard says, is that it requires our conscious participation. In the movement from hominid to human we were not conscious participants because our awareness was primitive and we were not capable of making reflective choices. Now we possess that terrible capacity, and it is essential that we engage it intentionally so that we may choose well. We must live up to the name "wise ones" if we are to continue to live at all.

There now exists an urgency to engage our consciousness that we have never experienced before. Our existence as a species depends on how we respond to the many issues that threaten the survival of Earth. Materialism, political egocentrism, religious fanaticism, and human ignorance have spawned a political and environmental crises of epic proportion. The crises involve every relationship on the planet: person to person, person to creature, person to organic and inorganic life. Nothing is exempt from the dangers that confront us. Sooner than later we will have to answer for our choices to move these issues to the fringes of our consciousness or to ignore them altogether. Our soil and water and air together with our suffering and oppressed sisters and brothers of all species cry out for our conscious, loving awareness.

As urgent as the current crisis is, however, we must nevertheless respond deliberately, with care rather than anger, with wisdom rather than fear. The quality of the consciousness with which we respond is as significant as the response itself, for in our responding we are becoming—we are taking the next evolutionary step toward—*Homo universalis.*

Hubbard describes the universal human as "one who is connected through the heart to the whole of life, attuned to the deeper intelligence of nature, and called forth irresistibly by spirit to creatively express his or her gifts in the evolution of self and the world."[7] These words can also describe those who practice some form of contemplative prayer or meditation. Fidelity to a silence that penetrates the heart and soul enables us to connect to the whole of life with hearts filled with peace, wisdom, and compassion for all that is. Out of this place we see more clearly what is and can respond with more awareness as we make choices regarding what can be. Out of this place, seeing without the blinders of illusion, we are radically amazed by what is around us.

This radical amazement brings us to our knees in gratitude and simple awe, then propels us into action that is informed by that gratitude and the certainty that the fire of the Spirit is alive in us. In short, we engage in acts of love. We are able to do so because Love itself fashioned us. We love, participating in the love that is all around us. We love, because love is the only appropriate response to the crisis that stares us in the face and challenges us to do what we must. We love because the Spirit that is Love takes up residence in us and uses us as a base of operations in our tiny part of the universe. But this love in which we participate is not a soft, fluffy feeling that cushions us in comfort. It is a grit-filled grace that enables us to make difficult decisions in the face of a sometimes desperate reality, choices that either spring us forward in the evolutionary process or threaten to do us in as a species.

Evolution, rather than degrading or diminishing humanity, asserts that we are part of a magnificent whole that consists of a vast web of relationships, an organic and cosmic body that evokes pure wonder. Yet, with all the knowledge that has accumulated, with all the fields of study that offer explanations that help us catch a glimpse of our profound connectedness, we still must come to the brink of Mystery. The theory of evolution merely offers "how" we have come this far. Only Mystery can provide a "why" and point us toward our purpose and meaning.

The fact that we are the universe come to conscious self-awareness is radically amazing. Consider this: You are able to sit and read these words as the result of 13.7 billion years of development. Over the last four billion years, life on Earth has developed from primordial cells that did not even have a nucleus into *Homo sapiens*. The water in your body contains primordial hydrogen formed in the first seconds of the Big Bang. The carbon atoms that form you came together as a result of the explosion of a supernova. The concentration of salt in your body matches the concentration of salt in the ancient seas. Your cells are direct descendants of unicellular organisms that developed billions of years ago. You have a reptilian brain and are able to walk courtesy of vertebra that developed 510 million years ago. You see because chlorophyll molecules mutated so that, like plant leaves, your eyes can capture the light from the Sun. And in your mother's womb your tiny body repeated the whole process of multicellular life on Earth, beginning as a single cell and then developing greater and greater complexity.[8] Caroline Webb expresses the heart of evolution and the perspective it can grant to us:

> Our bodies express all the history of life on this planet. And that history is also the history of every mountain, every river, every ocean, every pond and every millimeter of rock and soil, every wisp of water vapor and every breath of the atmosphere blowing ceaselessly around our globe. In our bodies flows the knowledge of an entire planet, an entire solar system, and the universe. It is nothing short of spectacular! What cause for celebration![9]

Celebration indeed! Considering how intricately complex our bodies are, how can we do anything but cry out, "How marvelous are your works, O God!" Each one of us, no matter what species we are, no matter how small we are in the vast scope of creation, is a marvel, fashioned patiently and lovingly by the Holy Energy we call God.

Together the "how" of evolution and the "why" of Mystery give us an image of who we are and what we are about. We are the universe conscious of itself, and we are about using our consciousness to reflect, to know, and to make choices in behalf of the whole. Rather than being a process that degrades human status, evolution allows us to glimpse a profound mystery, one that reverberates all the way back to the ancient mystics who say, "We are all one."

WE COME TO CONTEMPLATION

Imagine life after the Big Bang. Imagine that you can see microscopically, that you can notice that out of a patch of inorganic matter you are watching, there comes a tiny indication of movement, an almost infinitesimal emergence of life. . . . Watch as eons of development pass before your eyes—atoms evolve into molecules, molecules into cell, cells into multicellular organisms. . . . Observe as the gift of respiration is introduced and organisms begin to breathe. . . . See the development of specialized cells that capture light and then develop further into chlorophyll cells that feed plants in the process known as photosynthesis. . . . Notice how vertebrae begin to develop, then appendages that give mobility . . . watch the development of the primitive brain, then see it evolve in complexity. . . . See this primordial process continue, one tiny step at a time, until a human being emerges—a creature whose development rests on and depends on all the development that went before. . . . See that human being become you, and sense how long it has taken the universe to create you—the very unique you who lives and moves and has being in the Mystery of Love.

* *What in this chapter do you find challenging? What do you find radically amazing?*

* *Sit quietly for a moment. What does the Spirit want you to see? How do you respond?*

W E   P R A Y

Divine Creator, how astounded we are to know how carefully and creatively you have fashioned us. Your patient tending began so long ago and continues in this very moment. Truly we are blessed. Help us to grow in awareness of how we are part of the Body of Creation, how we are connected to all that has been, all that is, all that will be. Help us to be conscious of Life, to understand how significant our role is, and to participate in your work by becoming co-creators with you. May we make choices that reflect care for our home in the universe, allowing your love to be expressed everywhere and to all. Amen.

CHAPTER SIX

I AM THE LIGHT OF THE WORLD: INCARNATION AND PHOTOSYNTHESIS

Again Jesus spoke to them, saying "I am the light of the world. Whoever follows me will never walk in darkness but will have the light of life."

* JOHN 8:12

Christ is the reflection of God's glory and the exact imprint of God's very being, and he sustains all things by his powerful word.

* HEBREWS 1:3

Jesus is unlike anyone else, however, in that in him we find a radical and complete openness to God's self-giving in grace. In this one product of evolutionary history, the cosmos accepts God in a definitive and absolute way.

* DENIS EDWARDS

Earth was born 4.45 billion years ago, an offspring of the Sun. For nearly half a billion years the planet's major activity was simply to cool, forming a hard outer crust that would continue to develop and eventually form mountains and valleys. Half a billion years later, life on Earth began to emerge in the form of prokaryotes—primitive cells without a discrete nucleus. What a radically amazing moment this was! For billions of years there was nothing but fire and stardust, and then, without fanfare, a tiny, lifeless speck of matter became alive! And not only did matter come to life, but it began right away to interact within and among itself. Life came accompanied by relationships, primitive, to be sure, but a fundamental expression of the connectedness that has been part and parcel of all life that has followed.

The theory of evolution maintains that mutations occur which alter the capacity of a life form in some way. If the alteration is beneficial, enhancing its life, the mutation is incorporated into the life of the form in an ongoing way. For complex organisms, the DNA is modified and the mutation is inherited by successive generations. From the very beginning the trajectory of the evolution of the universe has been toward life, evident as a journey of transcendence characterized by ever greater complexity.

One of the most awe-evoking moments in Earth's evolution came three billion years ago when a simple primitive cell mutated and began to capture light from the Sun in a process we call photosynthesis. The process is a chemical reaction that involves relationships among components, an interaction upon which virtually all life now depends. Through photosynthesis, tiny chlorophyll molecules are able to capture solar energy in the form of light and convert it into food and oxygen. The equation that describes the process is

$$(\text{Light Energy}) + 6H_2O + 6CO_2 \longrightarrow C_6H_{12}O_6 + 6O_2$$

In the presence of light add six molecules of water (H_2O) to six molecules of carbon dioxide (CO_2). The result is one molecule of sugar ($C_6H_{12}O_6$) and six molecules of oxygen (O_2). The process works something like this: Photons (light energy) from the Sun strike chlorophyll. As it absorbs the light, the chlorophyll, using electrons from water, is excited to a higher energy state. The water splits, releasing oxygen. The energy from the interaction is converted to adenosine triphosphate (ATP), a sugar which becomes food both for the plant and those who consume the plant.

Besides light from the Sun and water from the Earth, photosynthesis requires carbon dioxide, a gas that in sufficient quantities is toxic to humans and other complex forms of life. In the process of photosynthesis, carbon dioxide is converted to oxygen, which is released into the atmosphere. Virtually all of the oxygen we breathe is thought to result from photosynthesis. (This is the reason for alarm in response to the destruction of woods and forests. They are the lungs of the Earth, and our own respiration is intricately connected to hers.) As a result, almost all life on Earth depends on the tiny yet extraordinary mutation that occurred three billion years ago.

It is important to consider that prior to this evolutionary step—the step where the first cell developed the ability to capture solar light and convert it to food and oxygen—the Sun was constantly radiating its light toward Earth. Throughout her entire history our planet was bathed in sunlight, but until life evolved to this new place, the light was incapable of being the nourishing source that it is now. The relationship between Sun and Earth was not restricted because Sun could not give, but because Earth could not receive.

Each second the Sun converts four million tons of its mass into energy in the form of light. Four million tons of Sun is given over each second, never to be recaptured, and, as a result, we live. The generosity of the Sun is constant, bathing the Earth constantly in light. Like a loving parent,

RADICAL AMAZEMENT

she endlessly gives not only all that she has but all that she is so that her offspring may flourish. Photosynthesis is Earth's way of receiving the offer, and her response creates an intimate bond between herself and the Sun.

As I reflect on photosynthesis, I am drawn to consider the Incarnation—that definitive event in Christianity in which divine life spilled over into human life in the person of Jesus of Nazareth. Looking at the Incarnation through the dual lenses of evolution and the interaction of photosynthesis offers a new clarity about who Jesus is and his meaning for all creation.

Since time began the Holy One has been radiating light toward Earth in one continuous act of grace. Karl Rahner calls this grace God's self-communication, an outpouring of the very being of God that not only permeates but maintains our life. God's self-communication began with the Big Bang 13.7 billion years ago. It was in and through God's grace that stars were born and galaxies were formed. The Holy One was present as supernova explosions scattered bits of cosmic matter throughout the universe and forces like gravity began to draw matter together in relationship. God was not identical to these developments, but neither was God separate from them.

Divine self-communication continued as Earth entered into orbit around the Sun and took shape. God's grace was present, operating from within each creative advance, pushing toward life moment by moment, era by era. Through grace came plant and sea life, amphibians and winged creatures, reptiles and mammals. Unimaginable variety was flung out in an astonishing array: trilobites, fishes, sharks, insects, frogs, dinosaurs, whales, monkeys, cats, apes, grass, antelopes, elephants, giraffes, lions, hippopotami, dogs, camels, tigers, flowers, bears, pigs, cows, sheep, and wolves. If the creative work of God were to stop here, the awe and beauty would have been enough. But the Creator did not stop. The urge of the universe toward life was a manifestation of the Holy One's creativity, and that creativity would not be contained. At some point, about four million years ago, a small mammal stood erect. With a brain capacity a bit larger than that of a chimpanzee, this hominid was *Homo sapiens'* distant but direct ancestor.

Still, divine grace continued to work from within the evolutionary process. As hominids developed, their brain capacity increased and they were characterized by eyes with a frontal focus and hands that, thanks to the ability to stand upright and walk on two feet, were freed to grasp and direct attention to objects. Grace-filled evolution pressed on, until some 2.6 million years ago, the first species that we recognize as human, the first member of the genus *Homo* (human), emerged. And just as photosynthesis was a breakthrough moment in the relationship between Earth and the Sun, so the emergence of the first human species was a breakthrough moment in the living world's relationship with the divine. Then 2.4 million years later, just about 150,000 years ago, our species emerged. As far as we know, for the first time in the 13.7 billion year history of the universe there was the kind of consciousness that we call self-awareness. Unlike other species, *Homo sapiens* had the capacity to know that they knew. *Homo sapiens* could reflect upon their own experience and begin to ask questions about their origin, meaning, and purpose. In evolutionary terms, the universe had become conscious of itself! How radically amazing!

About 100,000 years ago this new species developed the capacity for language, the ability to communicate in symbols. Words are symbols that represent some aspect of the communicator's experience. This development marked another significant moment in the story of the universe, since it enabled human beings to exchange information about events or entities that were not immediately present. Language, symbol flowing out of conscious memory, was to set the species apart from all others in a definitive way.

Before long the new species made the connection that there was something bigger than themselves, some Other whose power could be witnessed throughout nature in the

RADICAL AMAZEMENT

Sun and moon and stars, in thunder and rain, in fire and water. Even the earliest humans knew intuitively that there was some mysterious force or forces intricately involved in their lives. They called these mysterious forces "gods." Whether believing in one god or many, groups of humans began to use the capacity for language to create myths and rituals in order to explain and celebrate this conscious connection with the divine Other.

Now for 13.7 billion years God's self-communication had been radiating toward Earth and eventually on Earth as life evolved into greater and greater complexity. It was indeed a breakthrough moment when one species became self-aware and received in a conscious way the radiant grace that had always been present.

Our religious tradition is rooted in our own species' ability to use language to reflect upon experience and grasp intuitively that we are not alone, that there is a gracious power or energy that is at the heart of all life. This divine force, the Creator who has always been present within all life, revealed itself in the conscious awareness of *Homo sapiens*. Our Hebrew Scriptures can be summed up as the story of how a group from the human species discovered that this gracious Energy wanted to be in relationship with them. Over and over again the Hebrew people grappled with who God was and how divine grace was to be lived out in a conscious way. Sometimes the image of God during this stage of human development seemed very much like an unpredictable force of nature (after all, the first notions about gods came from encounters in the natural world). The Creator hovered and brooded over her creatures like a mother. Yahweh was sometimes angry and vengeful and violent. The *Shekinah*, divine Spirit manifesting as a guiding presence, accompanied them in the desert. Psalmists sang to the glory and majesty of a God who could smite enemies and comfort the afflicted. All the while, the consciousness of the "people of God" grew as they grappled with who they were in relationship to the God who was self-communicating and self-revealing in a way they could comprehend.

A critical movement in that evolving understanding of God came in the person of Jesus of Nazareth. This first-century Palestinian Jew changed everything for us. He is the Incarnation, which means "God-in-flesh," the definitive revelation upon which Christianity rests. We must remember that Jesus was human—*Homo sapien*. We tend to forget that our doctrine proclaims that he is *fully human*, not God disguised as human. This human being is significant because he embodies the next major evolutionary step in human consciousness.

The process of photosynthesis provides an image to guide us. Recall that the Sun has always been radiating her light energy toward Earth, ceaselessly pouring out four million tons of herself each second. In one sense all that radiance was lost simply because there was no receptor until photosynthesis came along. It took eons to lay the groundwork required, but eventually Earth was ready and photosynthesis began. With the emergence of that first tiny cell able to absorb the photons and convert them to food and oxygen, the Sun's radiance was finally capable of being received. And since that event no life on Earth has ever been the same.

Just like sunlight, God's grace has always been radiating toward Earth, ceaselessly self-communicating, ceaselessly pushing for life from within and without. With Jesus comes the breakthrough moment. After eons of preparation, humankind is finally able to receive grace in a more conscious way. Through Jesus and his interaction with the Holy One, Light breaks through into life in a way never before experienced. Jesus is able to absorb the gracious radiance of God in a fashion that transforms those in his midst who are ready to receive the breakthrough event. Jesus is able to know the Holy One in a way that shatters eons of illusion. Using our species' capacity to communicate, Jesus began to express his knowledge in a radically amazing way. Denis Edwards writes that in the person of Jesus

God's self-communication to all creatures reaches its concrete and tangible expression in history. Here at one point in space and time, in one flesh and blood person, God's self-communication is both given irrevocably and accepted radically.[1]

The universe, developing in and through the love of the Creator in space and time over billions of years, has finally evolved to that place from which it can respond fully to the Creator in the person and symbol of Jesus. His radical acceptance of Creative Love completes the circle that began with God's self-communicated grace starting with the Big Bang, continuing through the birth of Earth and the development of life, leaping forward with the dawn of consciousness and emerging with the awareness, embodied in Jesus, that all life is accepted and included in God's love and grace.

What Jesus began to teach about the divine energy is that it is inclusive, that in its radiant presence it embraces all that is. The connectedness of all life is not merely a physical phenomenon but an essential expression of that divine presence. Like the Sun which pours out her own life to nourish and support life on Earth, the Holy One pours out its own life to nourish and support us. And the nature of this outpouring is love, a love that knows no boundaries to its sacrifice and no limits to its profusion. Just as chlorophyll, water, and carbon dioxide engage in relationship with the Sun's energy and are thereby transformed, so we are through Jesus invited to engage in relationship with the Gracious One, allowing ourselves to be transformed by Love into love. And as we engage the Holy Mystery, we reflect more of its true nature—we incorporate Light into our being, and begin to radiate in a way that nourishes those around us with the breath of the Spirit. We become compassion and service, wisdom and grace, an inclusive love that flows out of the experience of connectedness that is our essence.

Through his new capacity for engagement with the Holy One, Jesus paves the way for all human beings to receive Mystery in a radically new way. We profess that Jesus is the definitive revelation of God in human life.[2] Through him Love is revealed in greater depth, and as we connect to that depth, allowing it to nourish us, we connect to others in greater depth, and our lives transform.

Summer mornings, when I am able to work at home, I often begin the day on the patio, where the morning sunlight streams in. I like to close my eyes and absorb the sunshine, allowing its warmth and light to penetrate my skin. I sit in silence, aware that God's light is entering every cell of my body. Receiving the light, I imagine that I begin to radiate, pulsing with divine light, becoming a little more aware, a little more capable of being open, just as Jesus was. I imagine Jesus sitting under the Sun in Galilee, wondering if he was conscious of how much he radiated the Holy One's presence. With awe I realize the Sun that illumined him is the same Sun that shines on me and on each of us. The same radiant presence that pulsed through him now pulses through us. This is the gift of the Incarnation.

Jesus is an evolutionary step forward in the development of humankind, an extraordinary development that changes all subsequent life on Earth. But evolution does not stop. It is an ongoing process, something of which Jesus must have been aware when he said to his disciples, "Very truly, I tell you, the one who believes in me will also do the works that I do and, in fact, will do greater works than these, because I am going to the Father." (John 14:12). Because of Jesus' definitive connectedness with the Holy One, we are all empowered. We all have access to the divine energy that permeates the universe and continues to create. To be conscious of this, to contemplate it and live out of it day by day, is to become a co-creator, just as Jesus was, doing the things he did.

And more. We will do these works and more. As evolution continues, how will humankind's capacity to receive God's self-communication swell? As evolution continues, will our self-awareness increase to the point that we become a new species, *Homo universalis,* universal

RADICAL AMAZEMENT

humans with the capacity to integrate the tension between our unique self-expression and the needs of the Earth community, creatures with the capacity to live in radical awareness of our connection to the divine? We are on the cusp of an evolutionary breakthrough, one that requires our conscious participation in Radiant Love, one that requires us to participate as co-creative agents of Love itself, to do the works that characterized the life of Jesus— and more.

WE COME TO CONTEMPLATION

* *Imagine you are present 4.45 billion years ago, as Earth takes shape and begins to hang in orbit around the Sun. Witness the cooling process and the way hot gases and liquids begin to form an outer crust. . . . Watch in utter amazement as simple cells emerge and spring miraculously to life. . . . See life multiply under the radiant light of the Sun. . . . Focus on a particular cell that draws your attention, and gaze at this cell as something incredible begins to happen. It starts to absorb the light from the Sun and begins to engage in the chemical reaction called photosynthesis. . . . Marvel as this tiny cell gives off food and oxygen, knowing that you are witness to an evolutionary moment that is truly revolutionary. Because of what you have witnessed, life on Earth will never be the same.*

* *Move forward in time a couple of billion years and gaze at the array of flora and fauna that proliferates the land. There is no end to the beauty and grandeur of the spectacle before you. . . . See a small, seemingly insignificant mammal stand erect and begin to walk, and know that you are witnessing a truly giant step for humankind and creation generally. . . . Watch as this newly-evolved creature learns to grasp objects with its hands and focus attention on them. . . . Be*

astonished to see the development of language and the ability to communicate in symbols. . . . Watch as the species continues to flourish and develop the capacity for self-awareness. . . . See groups gather to tell stories around campfires and enact rituals that bear witness to their recognition of something greater than themselves. . . . And now enter first-century Palestine, to a region called Judea, and begin to listen as an itinerant preacher named Jesus begins to teach about a God whose nature is love and connection, sacrifice and compassion. . . . Finally, see yourself here, now. Like all these moments you have just visited in your imagination, you are a product of evolution. Consider that you are connected to all these past moments in a vital way, and yet are also on the cusp of something new. As you sit in the radiant love of the Holy One, open yourself to that love and ask for the courage to respond to God's invitation to evolve. . . .

* *Does the metaphor of photosynthesis change how you image Jesus and his mission? How? What new thoughts or feelings do you have?*

* *What in this chapter do you find challenging? What do you find radically amazing?*

* *Sit quietly for a moment. What does the Spirit want you to see? How do you respond?*

WE PRAY

Radiant Love, since time began you have been communicating yourself to all creation. From the initial flaring forth to the convergence of galaxies, from the distant stars to Earth that you have made to be our home, you have never ceased shaping and fashioning and urging us toward life. We are truly amazed at the work of your hands and for the life of Jesus, who continues to teach us how to receive your radiance. We are humbly blessed by the gift of self-awareness that you have given to human beings. Help us to be mindful that we are the universe conscious of itself and that this capacity is not for our benefit alone, but for the good of all creationkind. We thank you for ongoing creation and for your invitation for each one of us to be co-creators with you. Amen.

CHAPTER SEVEN

YOU ARE THE LIGHT OF THE WORLD: MORPHOGENIC FIELDS

The one who believes in me will also do the works that I do, and, in fact, will do greater works than these.

* JOHN 14:12

You are the light of the world. A city built on a hill cannot be hid. No one after lighting a lamp puts it under the bushel basket, but on the lampstand, and it gives light to all in the house. In the same way, let your light shine before others, so that they may see your good works and give glory to your Father in heaven.

* MATTHEW 5:14–16

Humanity alone is called to assist God. Humankind is called to co-create.

* HILDEGARD OF BINGEN

The divine nature of Jesus Christ is precisely the same divine nature which is creatively at work in all cosmic history. It is the power of self-transcendence at the heart of the universe. The risen Christ is radically and permanently one with the absolute being which empowers the universe.

＊ DENIS EDWARDS

While in John's gospel Jesus declares that he is the light of the world, in Matthew's gospel, Jesus proclaims "You are the light of the world." With that assertion and the promise that we will do the works he has done—and more—it seems that Jesus, the Word made flesh, did not view himself to be the Holy One's final word to creation. From an evolutionary perspective as Incarnation, Jesus embodied the radically amazing advance in consciousness that enabled humankind to respond fully and completely to the radiant love of the divine. In Jesus we experience God's grace "given irrevocably and accepted radically,"[1] completing the connection between the human and the divine in a way that affirms our human capacity to live in conscious, co-creative relationship with the Holy One whose definitive characteristic is love.

The revelation in and through Jesus was not intended to be an end in itself, but instead was meant to usher in a new age of connectedness that Jesus called the "Kingdom of God." This reign of God would unfold as the human species embraced a new way of living characterized by compassion and service, freedom and love. And from the actions of Jesus—healing the sick, dining with "sinners," and teaching about the depth of God's love—God's reign, rather than being a future reality in another place, was to be a lived experience in the here and now. This is the vision Jesus lived and died for, the vision he passed on to us.

Jesus' assertion that we will do greater works than he did seems to place on us a great deal of expectation. While Jesus never asked his disciples to do more than they were empowered by the Spirit to do, he always challenged them to do more than they had ever dreamed or imagined. And so if Jesus says we can and must do more, then the more must be possible. Later in the Bible we read that "divine power has granted to us all things that pertain to life and godliness" through Christ, that we may "become partakers of the divine nature" (2 Pt 1:3–4). Thomas Aquinas echoes this belief, "The Incarnation accomplished the following: that God became human and that humans became God and sharers in the divine nature."[2] It seems that the life of Jesus and all that he was about was never intended to be an isolated event in the history of the universe but rather an evolutionary advance that brought the entire human species to a new level of awareness. In Jesus God became human so that we, too, could claim our share in divinity.

In Jesus the Holy One activated an emergent potentiality that broke open, flared forth, and transformed the cosmos. But the light was not to be contained or confined to him alone. The tangible expression of the reign of God requires that those who believe in Jesus also believe in his affirmation that we, too, are the light of the world. We, too, are called to become light, partaking of the divine nature in concrete and tangible ways.

For some, talk of claiming our divinity seems scandalous, and in one sense it is, since *scandulum*, the Latin root of "scandal," means "to snare." But the snare here does not come from some "new age" claim that has no substance, but rather from our illusion about who God is and what it means for us to be made in God's image and likeness. So often we reflexively describe God with words such as "all-powerful," "all-knowing," "omnipotent," or "almighty." God, we think, is a supreme being who is categorically unlike us, so superlative in every way that we cannot conceivably have any resemblance.

While the God of all creation can be described in terms of power and might, Jesus' life and teachings emphasize

RADICAL AMAZEMENT

something quite different. Through Jesus—the one who definitively tells us what God is like—we learn that the Holy One is more compassionate than we can imagine: accepting the unacceptable, loving the unlovable, inclusive in hospitality, healing all who want to be whole. We learn from Jesus that God is more about serving than being served, that the Holy One chooses vulnerability over might and continuously surrenders power in the attempt to set us free. God is about forgiveness and healing, justice and mercy.

What is truly scandalous is our reluctance to affirm that this is who God is and that, being made in God's image and likeness, we do in fact embody some of the very capacities that Jesus tells us are fundamental to God's nature. In moments of contemplative pause when we are acutely aware, we may recognize a resonance and understand the call. We, too, are capable of compassion. We, too, can accept the unacceptable and love the unlovable. We know how to serve, how to forgive, how to be just and merciful. We can be inclusive in hospitality, vulnerable before love, and empowered as we break free. To become divine is to simply but fully live out of the truth of who we are in the deepest part of our being— acknowledging that alone we cannot do any of these things, but empowered by the Spirit, we can move mountains! And so our refusal to claim our godly nature, rather than being an act of profound humility, is a rejection of the empowerment that came through the life and death of Jesus. And that rejection has been deadly, not only for humankind, but for all creation.

Even if we affirm that we have a share in divine nature, the questions remain. How do we do this? How do we live in fidelity to the gift of divinity? How do we participate in the next evolutionary step and affirm even more deeply our essential connectedness to the Creator? How do we find the strength and wisdom in this time of travail, not merely to carry on but to carry out the challenge of transformation that is so critical now? After all, like Paul we so often say "I do not understand my own actions. For I do not do what I want, but I do the very thing I hate" (Rom 7:15). So often we feel essentially powerless over the course of our own personal lives, let alone the emergence of the universe!

Once again we turn to the work of contemporary science to provide an image that may be helpful. Biologist Rupert Sheldrake has developed a hypothesis known as Causative Formation, which postulates that morphogenic or morphogenetic fields (from the Greek, *morphe*, "form") are unseen forces that preserve the form of self-organizing systems. A self-organizing system is a form or structure that maintains itself from within. Examples of self-organizing systems are human beings, families, communities, organizations, the biosphere. An automobile, on the other hand, is not a self-organizing system, since it is composed of mechanical parts assembled and operated from outside itself by others. A self-organizing system is living and dynamic, preserving its internal integrity while evolving creatively in response to its environment. Since each system has its own morphogenic field, our world is filled with fields that overlap and interact. Morphogenic fields are not energy fields, but they help manage the energy of a system by carrying information that maintains its wholeness. Bede Griffiths explains:

> Although there are fields of energy in the universe, the universe cannot be explained in terms of energy alone: there also has to be formative power. This formative power exists as non-physical, non-energetic fields which Sheldrake calls "formative causes" or "morphogenetic fields."[3]

In the human species, morphogenic fields set up habits of thought, activity, and speech. In one sense they are memory fields. Rupert Sheldrake states that "each member of a species draws on the collective memory of the species, and tunes in to past members of the species, and in turn contributes to the further development of the species."[4] Over time these remembered habits become part of the internal code of the organism and help subsequent generations learn newly-emerging patterns. Again, Bede Griffiths clarifies the concept:

> As an organism starts to develop it begins to resonate to a certain field, and the more the organism follows that particular path the more it becomes habituated and goes on developing within that field to its final form.[5]

We know intuitively that living systems cannot be described in terms of their parts alone. A human being is more than just an aggregate of tissues. There is more to a community than a gathering of individual persons. The biosphere is not simply an assembly of a mountain here and a forest there. There is always some invisible, ineffable essence or presence that defies precise explanation. The morphogenic field, a force both in and around an organized whole, is Sheldrake's explanation of this invisible, ineffable presence.

In my reading about morphogenic fields I came across several interesting accounts that help give a clearer understanding of what they are. One story involves the Russian scientist Ivan Pavlov (1849–1936), familiar to us in his work with dogs and behavioral response.[6] In one laboratory experiment Pavlov trained a group of rats to run to a particular feeding station at the sound of a bell. The first generation of rats to receive the training required an average of three hundred tries before always running to the feeding station when the bell rang. The offspring of these first rats—without any training from their parents— required only one hundred tries before learning to always run to the station at the sound of the bell, and their offspring, the third generation, needed only thirty tries to learn the behavior. Unfortunately Pavlov died before being able to investigate the cause of the rats' behavior.

Follow-up studies were done, however, by other researchers. At Harvard, Dr. William McDougal performed similar experiments, and he, too, found that successive generations of rats—untaught by their predecessors—needed less time to learn a new behavior. By the thirtieth generation his rats mastered the new behavior in less than twenty tries, while the first generation had required more than 165 tries before learning the new behavior.

The story becomes more interesting. In an attempt to duplicate McDougal's experiment, biologist F.A.E. Crew, located in Edinburgh, Scotland, used rats that were genetically similar to those used in the Harvard study but which were not descended from them. What Crews found was that his rats could master the same new behavior in only twenty-five attempts. At the same time in Australia W.E. Agar replicated the experiment, and his rats also learned the new behavior in about twenty-five attempts. Agar continued with successive generations of rats, and by the fiftieth generation the number of attempts at learning the new behavior decreased even more. It seems that once behavior is learned it becomes easier for subsequent generations to learn what was initially a new and difficult behavior. This phenomenon also appears in humans, with the achievement of new of behaviors easier in successive generations, whether the learning involves cognitive skills such as the acquisition of language or physical skills such as riding a bicycle. These observations may be considered evidence that points to the existence of Sheldrake's morphogenic fields.

Experiments with humans are yielding interesting results as well. The Princeton Engineering Anomalies Research (PEAR)[7] project, which began in 1979 under the direction of Robert G. Jahn, engages in the scientific study of various types of consciousness. In 1998, Roger Nelson, now the coordinator of PEAR, initiated the Global Consciousness Project (sometimes called the EGG Project), which uses sophisticated devices called Random Event Generators (REGs) to measure various types of consciousness in groups of people. The REGs consisted of an electronic sensor to detect quantum tunneling occurring around it. These sensors then fed the "random" information gather into a computer. A number of REGs—also referred to as "eggs"—have been set up all over the world in an attempt to measure anomalies in group or global consciousness. The data collected is then analyzed to see if anything of statistical significance occurs in global consciousness when an unusual event transpires. One researcher in the field writes:

> Accumulated data show a persistent pattern: when we are collectively engaged by powerful events, the network of REG detectors responds with a tiny correlation. The odds are about a million to one that the overall result is not a chance outcome, but an indication of something like a global consciousness field.[8]

One of the more interesting pieces of data collected by the Global Consciousness Project relates to the events of September 11, 2001. On that day 37 "eggs" were reporting from all over the globe. Charts that recorded the current of random events over the hours preceding, during, and after the bombings of the World Trade Center show that there was a spike in the current being measured at precisely the moment the first airplane hit the building at 8:45 a.m., before media coverage was possible. The surge intensified when the second airplane struck the next tower at 9:03. By that time media reports could account for increased global consciousness. However, what the data also documents is that the rise in global awareness recorded by the REGs actually began at 4:00 a.m., nearly five hours before the first plane hit its target. In other words, global consciousness was somehow alerted as the hijackers began to execute their plan.[9] Although the scientists themselves acknowledge that their work is far from complete, the research results seems to indicate that morphogenic fields are a distinct possibility.

In another study, the National Demonstration Project to Reduce Violent Crime and Improve Government Effectiveness, scientists were able to measure the effects of meditation on violent crime. In 1993 approximately four thousand practitioners of meditation from more than sixty countries came to Washington, D.C., to conduct an experiment to see if their practice could affect that city's high crime rate. The team of scientists directing the study predicted a large reduction in crime, based on the results of 41 previous studies. From June 7 to July 30 participants meditated together in large groups twice each day with the intention of reducing stress and violent crime. Statistical analysis demonstrated an almost twenty-five percent drop in violent crime during those two months, a reduction that could not be attributed to any other factor.[10] One way to describe what happened is that the meditation of so many practitioners together created a morphogenic field that influenced both energy and behavior in a positive and lifegiving way.

To summarize, a morphogenic field is a force that operates both outside and within a self-organizing system to help it maintain its organization or form. Like gravity or magnetism, these force fields cannot be seen, but their influence can be experienced and measured. They do not consist of energy, but rather are information systems that preserve the integrity of the system. Force fields contain a kind of memory that provides coherence and enables successive generations to learn behaviors more easily. Eventually the novel becomes habitual, deposited in the collective memory of the species. Morphogenic fields preserve stability while allowing for creativity and the emergence of something new.

How can the idea of morphogenic fields help us claim the vision that Jesus passed on to us—the dream of the reign of God in the world, with us as its light? We must first recognize that as self-organizing systems, morphogenic fields are integral to the identity of each of us as individual persons and as members of various groups. The thoughts, behaviors, and words that come from us contribute to the quality of the fields in which we are maintained. In other words, as the universe having come to consciousness, we make choices that contribute to the

RADICAL AMAZEMENT

condition of our fields. And these fields have formative power, shaping and affecting the quality of the energy around us.

One of the first movements toward claiming our light—our divine nature—is to examine our own morphogenic field of behavior and attitudes, asking ourselves how closely our values align with those that Jesus taught. His parables—like the prodigal son, the sower and the seed, the lost coin—all expose the condition of our personal field and point us toward the light. Commandments such as "Love your enemies" (Lk 6:27) and "You shall love the Lord your God with all your heart, and with all your soul, and with all your strength, and with all your mind; and your neighbor as yourself" (Lk 10:27) take us to the heart of Jesus' vision. Paul is helpful, too:

> Lead a life worthy of the calling to which you have been called, with all humility and gentleness, with patience, bearing with one another in love, making every effort to maintain the unity of the Spirit in the bond of peace. (Eph 4:1–3)

Scripture tells us what our formative field is to be like. When we maintain ourselves as a system committed to love, we create a morphogenic field of peace, one in which gentleness and patience flow out of a humble recognition of our connectedness to all.

Besides being rooted in scripture, the capacity to operate out of our lived experience of Jesus Christ will keep his memory and mission alive, allowing his vision and values to become more deeply imbedded in our field of awareness. The keeping of the memory and mission of Jesus is not about making him the object of our worship but about making ourselves the locus of the reign of God. As we live the life to which we have been called, our habits alter the morphogenic field and we help create a hospitable environment that enables others to respond as well. They are pulled in, unable to resist the love that resonates around them. In community—as self-organizing systems—we maintain and creatively advance this particular kind of morphogenic field, and it becomes a formative power that contributes to the transformation of the universe.

Many years ago I had a dream that radically altered how I look at the world. In this dream I saw a honeycomb—sweet, succulent, dripping with nectar. No regular honeycomb, this one filled my field of vision and seemed at first to have no beginning or end. As I examined its golden brown beauty, the honeycomb began to change before my eyes. In some mysterious way I saw all the relationships in my life, each one connected to another and another, the connections stretching out endlessly in webs of relatedness. I realized that through each person I touch many others are touched, and together we touch one another all over Earth. The honeycomb became Earth herself, and I knew, with absolute conviction, that we are all one, that our nature is unity, even when we do not see it.

The image of the dripping honeycomb embodies that truth for me, creating a morphogenic field that continues to grow, continues to draw and be drawn by others who resonate with the experience of connectedness. The image helps me attend to the quality of my connections, knowing that I cannot choose not to be connected, but that I do have some say over the quality of my connections. What I trust is that all together we can do what I cannot possibly do alone—acknowledge our divine nature and respond as fully as possible to the invitation and command: "You are the light of the world." Clarissa Pinkola Estes expresses the message well:

> The light of the soul throws sparks, can send up flares, builds signal fires, causes proper matters to catch fire. To display the lantern of the soul in shadowy times like these—to be fierce and to show mercy toward others, both, are acts of immense bravery and greatest necessity. Struggling souls catch light from other souls who are fully lit and willing to show it.[11]

To be spark-throwers, to send up flares, to be fierce with fire—this is what our world so desperately needs from us. Let us together claim our light so that once again all creation knows its fundamental nature and deepest truth. What a radically amazing invitation we have received!

WE COME TO CONTEMPLATION

* *I imagine Jesus with you. He is filled with light, with fire. See the light radiate out from him and watch as it touches his disciples and friends, the sick and the suffering, the lonely and afraid. . . . As he touches each person, notice that they, too, become filled with light. . . . Now he touches you, and you feel yourself aflame. The fire burns away any impurities within and you feel transformed, whole. . . . This transformation has happened in the ones around you, too, and you see them become whole. Then Jesus invites each one there to live as light, to claim it as your nature and your truth. Together, with the others, receive the invitation and respond. . . . And now begin to touch others who come into your presence. Soon the light continues to grow until your surroundings are aglow with love. . . . Watch as Jesus withdraws, knowing that you and the others now carry God's light and together you can extend it to the ends of Earth. . . .*

* *Describe your own morphogenic field. What kind of force or energy do you transmit to others? What would you like to change about your morphogenic field? How might you go about doing that?*

* *What in this chapter do you find challenging? What do you find radically amazing?*

* *Sit quietly for a moment. What does the Spirit want you to see? How do you respond?*

WE PRAY

God of Light, we give thanks that you have asked us to be bearers of your light—sparks of your very self, illuminating the world and making love tangible. Forgive the ways in which we try to evade being who we are, for the times we resist your invitation out of false humility or fear. As we become flames of your love, may we recognize that we are not separate from you, but because of your will we are participants in the divine Mystery. This day may we be conscious of bringing light everywhere we go, building up a field of love that will help transform the world. Amen.

CHAPTER EIGHT

FOR FREEDOM CHRIST SET US FREE: THE THEORY OF HOLONS

For freedom Christ set us free. Stand firm, therefore, and do not submit again to a yoke of slavery. For you were called to freedom, brothers and sisters; But do not use this freedom as an opportunity for self-indulgence, but through love become slaves to one another.

* GALATIANS 5:1, 13

The human person is a centre of consciousness which is capable of infinite extension and as it grows it becomes more and more integrated with the whole complex of persons who make up humanity.

* BEDE GRIFFITHS

God has arranged everything in the universe in consideration of everything else.

* HILDEGARD OF BINGEN

In the spiritual life, freedom is for nothing other than love.

✶ GERALD MAY

Have you ever taken a class in which you were given a "pretest"—just to see how much you need to learn? I wonder how we would do with a pre-test the first time we encountered this passage from Galatians: "For _____ Christ set us free." Perhaps "freedom" would not be the first word we would choose. There are other words that make good sense: service, community, love, peace, compassion, virtue. But the word Paul uses is freedom. Our freedom is Christ's aim for us, Christ's call to us.

Any of us can say a number of things about freedom, but much of what we say echoes the basic tenets of Western culture and the individualistic society in which we live. Recently the word freedom has been used to mean political independence and the liberty to pursue personal goals and dreams without oppression. Although political freedom is a basic human right, it is only one facet of the kind of freedom for which Christ set us free.

One popular notion of freedom has to do with "doing my own thing" in a way that disregards connectedness to others and even connectedness to our own inner being. This kind of liberty that is independent of the whole is a superficial kind of freedom that can eventually bind us up and achieve the opposite of what it intends. For example, our desire for "financial freedom" frequently causes us to be entrapped by materialism or consumerism in a way that depletes all our energy and creativity. We become enslaved to the very thing we thought would make us secure. So often we fall prey to an illusion about freedom, initially attracted by its sparkle, only to be ensnared by its dark and deadly reality. More and more we are discovering that our concepts of this human capacity are narrow and confining. From the point of view of a universe that is growing and expanding, a superficial understanding of freedom is no longer sufficient.

The theory of holons allows us to talk about freedom in a way that goes beyond the superficial. Although it encompasses the physical sciences, this theory as presented here comes from the work of philosopher Ken Wilber and is a major component of his Integral Model which unites physical science, social and cultural history, psychology and other related fields into a coherent system.

According to the theory, all reality is composed of holons, or whole/parts. Wilber defines holons in this way:

> A *holon* is a whole that is a part of other wholes. For example, a whole atom is part of a whole molecule; a whole molecule is part of a whole cell; a whole cell is part of a whole organism. Or again, a whole letter is part of a whole word, which is part of a whole sentence, which is part of a whole paragraph, and so on. Reality is composed of neither wholes nor parts, but of whole/parts, or holons. Reality in all domains in basically composed of whole parts.[1]

Reality in all domains includes everything in the universe, everything on Earth, whether we speak biologically, psychologically, or sociologically. For example, atoms are wholes—distinct modes of being that maintain their own integrity. Yet atoms are not merely wholes unto themselves but are parts of molecules as well. In the same way, molecules are wholes with their own distinct mode of being, but they are also parts of more complex wholes. They, too, are holons.

Holons emerge into ever more complex forms. A cell is more complex than either a molecule or an atom, and yet a cell is composed of molecules and atoms and depends upon them for its own existence. Each holon is nested in another less complex holon, and each emergent whole has greater complexity than the wholes that are its parts. Each holon that emerges depends on all the other holons below it to maintain its own being. The organization of holons nested within more complex holons nested within more complex holons forms a kind of hierarchy of development, or holarchy. It is through holarchical development—holons nested in holons in greater and greater complexity—that all being is connected. According to Wilber, to destroy any given holon will destroy every holon above it. If we obliterate molecules, for example, cells, organs, and organisms will no longer exist.

The Physical Level
organism organism organism
organs organs organs organs organs
cells cells cells cells cells cells cells cells cells cells cells
molecules molecules molecules molecules molecules molecules molecules molecules
atoms atoms atoms atoms atoms atoms atoms atoms atoms atoms atoms atoms atoms

HOLONS DEVELOP WITH INCREASING COMPLEXITY
Each level of holons is nested in the one below it

The same theory can be applied to the organization of systems as well. Individual human beings come together to form social groupings—families, communities, institutions, nations—and groups join together to form more complex organizations. Changing a single member alters the whole group or organization because it shifts the composition of the group. Each member brings consciousness and presence that are essential components of the larger whole.

Whether we are considering the physiological level of being or consciousness itself, the theory of holons reveals how everything and everyone is connected to everything and everyone else on the most fundamental of levels. Each distinct mode of being is always part of a larger whole. The parts compose and therefore affect the whole; the whole holds together and therefore affects the parts. Every creature is a holon, Earth is a holon, our solar system is a holon—all wholes are part of something more complex, and all are contained in the one Universe that is home to all. The molecule depends on the atom, the cell depends on the molecule, and all depend on the stability of the interconnected system in order to thrive. Connectedness is a fundamental reality we cannot escape, and attention to the quality of our connections—the various wholes and parts of our lives—can radically alter our experience by altering the way we see.

The Four Capacities of Holons

There are four basic characteristics to every holon:

* self-preservation (need for agency);
* self-adaptation (need for communion);
* self-transcendence (movement toward greater complexity);
* self-dissolution (movement toward dissipation and death).

THE FOUR CAPACITIES OF HOLONS

SELF-TRANSCENDENCE

SELF-PRESERVATION
(AGENCY)

SELF-ADAPTATION
(COMMUNION)

SELF-DISSOLUTION

RADICAL AMAZEMENT

Every holon has the capacity and the need to maintain its own wholeness or internal integrity, to be an agent which acts as a whole. In human beings this element is expressed in our desire for individuality and autonomy. Being able to define ourselves as having distinct personalities with unique gifts is an essential human task. This is the core of what it means to engage in self-preservation or agency.

The self-preservation of any holon is balanced by its capacity and need for self-adaptation. All holons are parts of greater wholes and therefore must adapt and in some ways limit their individuality or agency in order to become a functional part of the whole. In humans this characteristic is expressed as the desire to be in relationship or communion with others. Of necessity we limit or direct our individuality or self-expression in ways that allow for a greater expression of the whole to which we belong. Every holon holds self-preservation and self-adaptation in creative tension. The more self-preservation (agency) a holon has, the less self-adaptation (communion) it has, and vice versa. How well the holon negotiates this tension will determine the quality of its growth.

In the normal course of development the tension between agency and communion mounts. These two capacities, incapable of flexing any more, reach an impasse. This moment in the life and development of the holon is ripe with opportunity, for in the tension lies the potential for a new level of growth or perhaps an entirely new holon. The stress lays the groundwork for self-transcendence, the third capacity of the holon.

Normally each holon has the capability to become more than it presently is. Each one has the capacity and need to transcend current limitations and move toward greater complexity. There is in every holon the drive toward self-transcendence, toward becoming more complex. The greater the complexity, the greater the consciousness. We see this need in all levels and arenas of human development. The adolescent is the result of the child's transcendence, and the adult comes out of the transcendent capacity of the adolescent. Even as adults we strive to become more capable in our chosen fields, more competent in our skills, more consciously aware as we grow in maturity. Each time we successfully negotiate the conflict between self-preservation and self-adaptation, agency and communion, we exercise the capacity for self-transcendence, becoming more conscious, more aware, more capable in some way than we were before.

The self-transcendent characteristic of the holon is held in tension with its conflicting capacity, that of self-dissolution. Every holon can move in the opposite direction of transcendence. Each being can deny conflict, resist change, or simply break down under stress. Everything eventually dies. Just as the holon learns to balance and negotiate the tension between agency and communion, it must balance and negotiate the tension between transcendence and dissolution, life and death.

The capacity for self-transcendence evokes reflection about freedom. As human holons, we struggle to negotiate the ever-present tension between self-preservation and self-adaptation. We are moved to express ourselves, to speak our truths, to make the contributions that only we can make. But we also experience the deep-seated need for communion, to live with attention and intention in regard to the relationships that define us just as clearly as the expression of our gifts do.

Inevitably self-preservation and self-adaptation run into conflict. Sometimes we give over too much of ourselves and lose touch with our inner core. At other times we retain our individuality so rigidly that it suffocates the experience of communion. Often the pull and tug is unconscious (although on some level we experience the anxiety), while at other times the struggle is the result of painfully conscious choices. However we get to this point, it is ripe with potential for self-transcendence—for a breakthrough in freedom that is Christ's vision and call.

RADICAL AMAZEMENT

Within each and every holon there is a spark of divine creativity and the power to develop into something new. In evolutionary terms, holons move toward greater complexity. In human psychological terms, we become less egocentric and more integrated. In spiritual terms, we open our hearts to the Spirit who is eager to work in and through us for the transformation of the world. Much of the self-transcendence that occurs in the universe happens as part of normal growth and development, but some of the movement toward becoming more must be chosen—and choosing requires that we are free.

As with any holon, the human holon's capacity for self-transcendence is held in tension with that of self-dissolution. We have bodies that fail, we suffer losses that diminish us, we have dyings that encroach upon our lives, we have egos that can give way to fear and lock us in self-absorption. Rather than viewing this element in our lives as an evil or a threat, it can be profoundly helpful in sharpening our consciousness and allowing us to grow in freedom.

Becoming free—Christ's aim for us—is never automatic and never easy, even when it is the desire of our own heart. Living in freedom means that we are sufficiently unfettered by fear to enter into the process of self-transcendence, cooperating with the creative Spirit that asks us to become more than we ever dream or imagine we can be.

Living in freedom requires that we recognize the connectedness that is a basic reality of our existence, that we are holons within holons within holons. All we do affects all the other wholes of which we are a part and all the other parts that make us a whole. Living in freedom in a conscious way means that we are always becoming part of a greater whole, trusting that the Creator is continuing to create, not only within us, but all around us. Living in freedom also means being content to be incomplete and unfinished—because "complete" and "finished" are irrelevant as we participate in a process that began billions of years ago and will continue for billions more. Living in the freedom for which Christ set us free means not judging—ourselves, first of all, and all others—for everything is incomplete, everything is trying its best to live in the tension between self-preservation and self-adaptation, transcendence and dissolution, love and fear.

The human capacity for freedom that allows us to cooperate with the process of transcendence is certainly amazing—but equally amazing is that all of life has the capacity to transcend, all life has at least a primitive, rudimentary capacity to become more, to participate more fully, to affect all other parts in a way that contributes to transformation. Freedom cannot and does not exist in isolation. The freedom of any part is connected to and influences the whole. The freedom of any whole is connected to and influences its various parts. And since human beings are but one part of a greater whole, our anthropocentric notions of freedom must go. No longer can the human say to the water, "I have no need of you" or to the rain forest "I have no desire for you." Because all Earth's resources are intricately woven into a seamless whole, the way we value and take care of each part equals the way we value and take care of ourselves.

It is for freedom that Christ set us free—freedom that is an act of going out of the self, going beyond all fixed categories, becoming part of a greater whole, with greater meaning and greater significance than we can experience alone. The freer we become, the more attentive we can be to the quality of our connections, aware of how our spirit touches all the parts of our lives, aware of how our spirit touches all others. To live in freedom is to live with attentiveness to the intricate connectedness of all that is— an awareness that is radically amazing!

WE COME TO CONTEMPLATION

* *Close your eyes and get in touch with your body, feeling its density, its length, and width. Gaze upon your internal organs, the muscles, bones, the nervous system. . . . Look more closely, and see a portion of tissue, and then individual cells, then molecules, then atoms. Consider the fact that at this level, you are more energy than solid matter, that there are no solid boundaries that separate you from what is "other." . . . In this moment, there is no distinction between yourself and the rest of creation; you are participating in and sharing one energy, the energy of a greater whole of which you are only a part. Allow this reality to sink in: You are a whole composed of many parts, yet part of an even greater whole, part of a mystery that embraces all life. From this place begin to think about freedom, how freedom from within this contemplative space is different from other ways you have thought about freedom before. . . .*

* *What are the parts that make you a whole—not only the physical, but feelings, beliefs, memories, etc.? What are the holons of which you are a part? You may want to draw or list all the holons that you can think of that are in your life.*

* *Reflect upon the gift of freedom you have been given. How do you use that freedom for yourself and others?*

* *What in this chapter do you find challenging? What do you find radically amazing?*

* *Sit quietly for a moment. What does the Spirit*

want you to see? How do you respond?

WE PRAY

O Gracious God, you have created each of us with the capacity for freedom. It is for our freedom that Christ lived and prayed and taught and died. Help us to move beyond our narrow, self-centered ideas of freedom and grasp what it truly means to be free: to live creatively with the tensions that are ours, to choose lifegiving responses the best way we know how, and to trust that the power to transcend helps us to do "the more" that Christ calls us to do. Amen.

CHAPTER NINE

SELF-DISSOLUTION AND BLACK HOLES

I do not understand my own actions. For I do not do what I want, but I do the very thing I hate.

∗ ROMANS 7:15

General relativity reveals the possibility of objects so dense that escape speed exceeds the speed of light. Such objects curve spacetime so much that not even light can escape them. For that reason they're called black holes.

∗ Richard Wolfson

Our journey in self-transcendence and freedom is never a free ride, but rather a costly movement that requires the sacrifice of what we have thought our lives to be. Growth in freedom demands the diminishment of egocentricity, our tendency to define ourselves through behavior based in fear rather than presence founded in love. Those who continue to evolve into greater self-transcendence understand that no price is too excessive, for to live in freedom, to live in the light, is the purpose for which we have been born.

The characteristic of the holon that is always in tension with our capacity for self-transcendence is that of self-dissolution. The reason freedom does not come automatically or easily is that we also experience a natural pull in the opposite direction. Resistance to difference, fear of change, even the normal disintegration of our bodies and minds are part of the experience of self-dissolution. We have no choice about whether or not we encounter its energy, but we do have some say over how we will respond in the encounter. Another piece of the new universe story will provide a useful image.

Perhaps nothing in the new cosmology is as intriguing as black holes. They make for fantastic science fiction and unnerving movies, conjuring up frightful images of rocket ships that travel at warp speed and challenge the limits of spacetime, venturing perilously close to a black hole then being sucked into a vast darkness never to be seen again. While some of these depictions are purely fiction, black holes are certainly amazing!

How often we have said, "What goes up must come down." This isn't necessarily so. If we throw an object upward fast enough it can escape Earth's gravitational pull and continue to move out into space forever. The necessary momentum needed for an object to exceed the force of gravity and continue on forever is called its escape speed (also called escape velocity). Earth's escape speed is a little more than seven miles per second.[1]

Einstein's General Theory of Relativity suggested that there could be some celestial objects so dense that nothing within their gravitational field could reach escape speed, even light, which travels at 186,000 miles per second. Scientists have since confirmed the existence of such objects, called black holes. Black holes form when giant stars die and their mass collapses into a distinct, gravitationally-dense point called a singularity. A black hole is not an object like a star which has a mass that can be clearly delineated. Rather, it is more like a region of spacetime without precise boundaries but with a definite effect.

The region of a black hole is defined by what is called the event horizon, which is the point of no return for anything that ventures past its boundary.

> No information about events occurring inside this distance can ever reach us. The event horizon can be said to mark the surface of the black hole, although in truth the black hole is the singularity in the center of the event horizon sphere. Unable to withstand the pull of gravity, all material is crushed until it becomes a point of infinite density occupying virtually no space. This point is known as the singularity. Every black hole has a singularity at its center.[2]

The more massive the black hole, the larger its event horizon or sphere of influence will be.

Since a black hole is not visible, evidence for its existence comes from measuring the effect it has on surrounding matter. As particles are pulled toward a black hole, they gain speed and begin to reach very high temperatures, emitting X-rays that can be detected. There is strong evidence to suggest that at the center of every galaxy there is a black hole, including our own Milky Way.

Black holes vary in size. There are stellar mass black holes, relatively small ones that form as the result of the explosion of a comparatively small star. These have mass a few to ten times greater than that of our sun. Medium-sized black holes have the solar mass from a few hundred to a few thousands of our Sun. Supermassive black holes that lie at the center of most galaxies have a solar mass a few million to hundreds of billions times that of our Sun.[3]

What would happen to us if we did travel toward a black hole? Suppose we were traveling in a space ship and our X-ray equipment detects a black hole in the region. Our exploration confirms that there is no definitive boundary to the black hole, but we do notice that the gravitational pull is exponentially greater than what we have experienced on Earth. As we continue to move toward the black hole, the light around us grows redder, evidence of its loss of energy. Once we pass the event horizon—the point of no return—the gravitational pull becomes so great that space and time become distorted and will soon cease to exist. As we approach the black hole, our bodies are stretched further and further apart until we resemble a bizarrely long string of spaghetti. At some point we are drawn completely into the darkness, becoming one with the super-dense singularity.

The image of the black hole has found its way into our everyday vocabulary. Falling into a black hole signifies becoming pulled in to a social situation from which escape is difficult, even impossible. Military spending, for example, has been described as causing a black hole of debt from which even our children may not escape. We can have black hole experiences in every arena of our lives, situations or events in which we seem manipulated by unseen forces and which fall outside the accustomed gravitational pull that keeps us grounded.

We can experience black holes within ourselves. Sometimes the expected tension between self-preservation and self-adaptation can pull us toward a darkness that seems to threaten annihilation of one sort or another. There are also those areas of our personality or character that seem to set us on a collision course with disaster. Some of our habits, attitudes, and behaviors can fall into this category. We can become addicted to just about anything, from substances to points of view that exert increasing pressure and pull us away from what gives life. Unaware, we can travel into regions that grow progressively darker, where the gravitational pull becomes so intense that our integrity threatens to collapse. These encounters always jeopardize our freedom and challenge our capacity for transcendence.

The journey into this sort of black hole is similar to our imaginary travel into the farthest reaches of outer space. Sometimes we reach escape speed and break loose from gravity through our own doing. Sometimes life itself propels us into unfamiliar space. Whatever the cause, we find ourselves ungrounded, on a journey from which there is no chance of escape and into a place where there is no detectable light. Some black hole experiences even threaten the annihilation of who we are.

Like travel to a black hole, this journey of darkness that we sometimes make crosses no readily defined boundary, but is characterized by an ambiguous event horizon that quickly throws us off our bearings and takes us to a place of no return. Once we pass its indistinct perimeter all that we are becomes oriented toward the darkness. Familiar objects, no longer able to retain their shape under the intense pressure, become distorted and stretched beyond recognition. Cut off from ordinary experience, time and space stand still as we collapse in upon ourselves. We are utterly oriented toward a dense singularity that will not allow escape from its gravitational pull. All light dims, then vanishes completely as any trace of our former self disappears.

What is it that brings us to these black hole experiences? What is it that causes us to nearly self-destruct in spite of our desire to live as beings oriented toward the light? The apostle Paul seems to understand the experience when he says, "I do not understand my own actions. For I do not do what I want, but I do the very thing I hate" (Rom 7:15). Against his deepest desires, Paul seems unable to be as free as he can be, as loving as he longs to be. We have named Paul's problem sin, for indeed it is a "missing the mark" that causes him to struggle against an inner darkness that wants to suck the life out of him. We hear his anguish and identify with it all too well.

RADICAL AMAZEMENT

But sin is defined as "the purposeful disobedience of a creature to the known will of God,"[4] and perhaps is not the best word to use in our discussion of this all too human dilemma. In my experience many people are purposefully seeking to be obedient to the will of God, intentionally striving to live in the light, creating a field of life and love wherever they go. The difficulty is not with our intention but with our attention.

While some black hole experiences simply occur in the normal course of life, others come about because we are living unaware. When we are unaware, we adapt attitudes and habits that draw us closer and closer to darkness. Unaware, we make choices to relieve ourselves of reality, thinking to lessen our psychological pain or social discomfort. Before we realize what is happening we encounter a gravitational pull that exerts greater and greater force. No longer grounded, time and space become distorted as we orient toward the darkness and lose our self-control.

Everyone, it seems, has had experience with this kind of darkness. Most human beings know what it is to have been under the influence of a menacing force that threatens to break them apart. I think these occasions are normal and necessary, even though they are painful. We all have times when we are swept away, pulled out of our customary comfort and into places where we did not plan to go. Sometimes these journeys come because our own behaviors have thrown us out of orbit. At other times they come simply as the result of life and the chaos that always threatens to encroach. Usually we are halfway to the hole before we realize we are off course. Sometimes we skirt the perimeters without major damage, while at other times we come undone. Perhaps, rather than an anomaly, at least some black hole experiences are an essential part of everyone's journey, and what is important is to learn how to navigate the space in a way that draws on our capacity for self-transcendence.

One of the most important psychological tasks that we perform as adults is integration of the shadow, that unconscious part of our personality of which we are unaware. The term "shadow" comes from Carl Jung, who used the expression to refer to the whole of our unconscious personality. The shadow contains all of the parts of ourselves that we have repressed for the sake of our ego ideal. We want people to think we are congenial, for example, so we relegate all our unsocial inclinations to our shadow. In order to project a persona that helps us adapt to society and get our wants and needs met, we stuff all kinds of emotions and attitudes underground, hidden away in areas of our psyche that we prefer not to acknowledge. Especially at midlife these denied fragments of our unconscious begin to surface. We can find ourselves angry or depressed. We can discover that we hate what we once loved. If we happen to be living with contemplative awareness, we can often acknowledge and catch these movements and bring them into the light of love. But by definition the elements of the shadow are unconscious, and we are most often unaware of their presence until they pop up and propel us into some kind of darkness at warp speed.

Most of our personal black hole experiences come as a result of our failure to integrate pieces of our shadow. Conversely, integration of shadow pieces as they surface is evidence of self-transcendence. Since having a shadow is a given, our task is not to eliminate or judge it, but to strive to attend to it so that we grow in awareness of the whole of ourselves, not just selected fragments. Black hole moments are crises, to be sure. They will always lead to death of some part of us, but they also lead to possibilities and potentialities that we did not know were ours.

The Chinese ideogram for crisis is comprised of two characters, "danger" and "opportunity." A black hole crisis, whether of our own making or imposed upon us by circumstances outside our control, is a time of danger. We can choose to ignore the signs and become oblivious to the increasing gravitational force that pulls us into bondage. Addictions can propel us toward disaster and we may need the hope of a rescue mission, even when we resist it. We can become subject to the forces of self-dissolution and find all that we have striven for disintegrating around us.

But a crisis is also a time of opportunity, a place from which new choices can be made. Black hole experiences that result from the tension between self-preservation and self-adaptation can be resolved with creativity. Black hole experiences that come from our own choices can lead us to a new level of freedom. Black hole experiences that expose our need for healing can propel us into the quest for wholeness. Black hole experiences that bring us to the brink of death itself can affirm how gifted we are to be alive, to be expressions of the very Light that creates the cosmos.

It was once thought that nothing could get out of a black hole, but now, as a result of the work of Steven Hawking, we know there is at least the theoretical possibility of escape. A phenomenon called "Hawking radiation" implies that under certain conditions black holes can emit radiation. The theory says that the release of radiation comes from the creation of pairs of subatomic particles in the space adjacent to the black hole, with one particle getting pulled into the black hole while the other radiates away. This action is thought to potentially cause the eventual collapse of the black hole.

Another name for radiation is light. What Hawking radiation suggests is that not even a black hole lies outside the influence of light. Even the smallest of particles can reverse a situation from which there seems to be no escape. A black hole's darkness is not so definitive after all.

Does this not give hope to each of us who has fallen into the pull of a personal black hole? In every experience, even those in which the black hole has sucked us in and seemed to have annihilated all that we have known ourselves to be, there is the possibility of light. Light, we know, is the primary expression of Creation. It is Love's primary speech, the word made energy and Mystery made tangible. In our times of utter darkness, when all hope seems to have vanished, along comes the faintest particle of light that collapses the hole and makes us whole! Just like the darkness itself, the light that becomes a saving grace is nearly undetectable. It may be faint, but it is not feeble. It may not register on the visible spectrum, but it is nevertheless present and potent. And because it is there, dissolution will never have the final word.

What is required of us as we recognize that our efforts to transcend will always be countered with the threat of dissolution? The problem, I have suggested, is not our intention but our attention. But questions remain: What is it that will keep me attentive? How can I grow in awareness so that I may avoid unnecessary black holes and navigate the required ones with grace? We know the answer: contemplation.

Contemplation as a prayer practice slows us down so that we become aware that something is going on within us—something different, unusual, discomforting. In the silence we more readily recognize our resistance and fear. As we grow in fidelity to contemplative prayer, we begin to change the way we meet the world. We are more awakened to life, less needy, more clear about who we are and what our life is about. We walk in peace even when we are uncertain, in assurance even when all is dark.

In the experience of contemplation we avail ourselves to the creative power of the Light. We discover in the darkest of holes the presence of the divine. We escape from the distortion of spacetime and, no longer trapped in the past or the future, return to the present moment, to the now. Powerless to change circumstances on our own, we are empowered by grace to become grounded again and allow all things to work together for good—our own and that of the radically amazing cosmos in which we live.

WE COME TO CONTEMPLATION

* *Imagine that you are on a space ship traveling in the vicinity of a black hole. You cannot see the black hole, but you know that it is there. Experience the light around you shifting toward red, a sign that already the light energy is affected by the immense gravitational pull. . . . Feel the pressure begin to increase as your ship becomes*

drawn to the black hole. . . . Know that you have passed the point of no return. Feel yourself being stretched, stretched to the point of breaking. . . . Then know that you have dissipated into the darkness and been drawn into the singularity from which there is no escape. . . . And now be surprised by grace. Envision a single particle of light breaking through the darkness and annihilating the black hole. . . . See the emergence of the light that you never thought you would see again. . . .

* *Consider a time in which you had a black hole experience in your own life, a time when darkness seemed on the verge of doing you in. Was this a darkness of your own making, or was it thrust upon you? What was your response? Recall the experience of darkness as it engulfed you. What helped you escape? What was the rescue mission that saved you? When did light begin to break through for you again?*

* *What in this chapter do you find challenging? What do you find radically amazing?*

* *Sit quietly for a moment. What does the Spirit want you to see? How do you respond?*

WE PRAY

Jesus, as the human being who transcended even death, we know you understand our struggles with self-dissolution. You know that sometimes we have painful experiences because of our own failures, while at other times life itself seems to propel us toward black holes that threaten to devour us. We are grateful that our experience of divine grace assures us of a Light that will always overcome darkness, a Love that will always save us from fear. May we be faithful to our own contemplative path, our hearts ever more open to your transforming presence that sets us free. Amen.

CHAPTER TEN

DEATH AND RESURRECTION: SUPERNOVAS

But God raised him up, having freed him from death, because it was impossible for him to be held in its power.

* ACTS 2:24

The supernova is a multivalent event: simultaneously a profound destruction and yet an exuberant creativity.

* THOMAS BERRY AND BRIAN SWIMME

Jesus Christ in his life and death, in his grace-filled human reality, has become a power shaping the whole cosmos.

* DENIS EDWARDS

Death is not the opposite of life. Life has no opposite.

* ECKHART TOLLE

I t is inevitable. Despite all our modern technology and efforts to reverse or stall its coming, each of us has an appointed encounter with death. Nothing in the universe escapes the finality of its clutches. Death will come to us all.

In the previous chapter we used the metaphor of black holes to discuss the kind of darkness that is associated with the shadow aspects of our personalities. In this chapter we will be discussing another aspect of darkness, death itself, and the various ways it comes not only with our physical demise, but also including those kinds of deaths that encroach upon us and demand a yielding of ourselves in a way that is absolute and final.

We live in a culture that fears death and suffers great discomfort in its presence. We are not even very good with the little dyings—self-denial, delayed gratification, commitment to stay the course—that encroach upon our lives and offer the opportunity to build character and define values. Perhaps this is the expected outcome of living out of "me" rather than "we." Having lost the big picture, it is easy to grip tightly the little things that, in the long run, are of no consequence.

As I write these words it is Easter Monday, a quiet day after a week of symbol-filled liturgies that invite reflection on the mystery of life and death. How intricately each is bound to the other! How dramatically each teaches us about the other! While our liturgies separate the various strands of the Paschal Mystery, in truth we celebrate one event, one moment in time in which life/death has been encountered with unique consciousness and freedom. While death was nothing new and continues to be an integral part of life, Jesus' response to death remains significant, challenging us to consider how we will bring consciousness and freedom to our own death—and the living that will fill in the time between now and then.

Death is integral to life. It is the pull and tug of self-dissolution that strengthens our capacity for self-transcendence. It lulls us out of lethargy and moves us to name our desires and account for our time. Death's breath on the back of our necks can cause us to inhale life with new vigor. In this context it can become a consort with the creative Spirit that seeks expression in and through us. In quiet moments, with our fear settled at a distance, we come to recognize that creativity and destruction, life and death, fear and desire are all intertwined in a Mystery that is overwhelming in its intensity and power, yet is merely a wisp away. From this place we recognize that our attempts to escape the certainty of death pull us away from the reality that Christ taught us to embrace.

Death has always been a part of the ongoing development of the universe, and it is the image of supernovas that can lead us into reflection on this experience that poses such difficulty for humankind. Supernovas are the death eruptions of stars. The Hubble Space Telescope photographs of them are so stunningly beautiful that it is easy to forget that what we see are really images of death and destruction. They are simultaneously images of passing away and coming to be, icons that lead us into the heart of Mystery itself.

Three hundred million years after the Big Bang, the first stars and galaxies came into existence. Stars are flaming gaseous spheres formed by the collision of hydrogen with helium. Eventually, however, the fire depletes the hydrogen that fuels it and burns out. The core of the star eventually begins to shrink, growing hotter and denser, setting off a series of nuclear reactions that forestall its collapse. Eventually the core of the star turns to iron and there is no material left for fusion. The temperature at the core rises to over one hundred billion degrees as the star enters its final throes of existence. In the extreme heat the iron atoms are crushed together and they create a repulsive force that overcomes gravity. The result is a supernova explosion that decimates the star and flings bits and pieces of its former self out into space.[1]

RADICAL AMAZEMENT

Supernova explosions are uncommon. It is estimated that in the past one thousand years there have been fewer than five supernovas in the Milky Way Galaxy. On February 24, 1987, however, a supernova was detected in the Large Magellanic Cloud about 160,000 light years away, and the Hubble Space Telescope has been able to capture images of its incredible magnificence. It is thought that the star that exploded and gave off all this light had a mass twenty times that of our Sun.[2]

A supernova significant for our future existence occurred about five billion years ago in one of the spiral arms of the Milky Way Galaxy. The spectacular, fiery death of this star had repercussions that rippled throughout the galactic neighborhood. The light emitted from the explosion, though short-lived, was brighter than that of a billion stars. What was once a giant stellar object was reduced to bits of cosmic debris.

Elements from this fiery eruption collided with a nearby cloud of hydrogen gas and interstellar dust. Tremendously hot fragments began to come together under the influence of gravity, and a new star began to form. We call this star the Sun. As she cooled and began a regular rotation, small cosmic objects called planets began to orbit around the Sun, and in another billion years simple life forms began to appear on the planet Earth. Just over two million years ago—the blink of an eye, cosmically speaking—there was the dawn of a life form that could acknowledge the Sun's existence and begin to reflect upon her meaning in their lives. These were the first humans.

Life on Earth is nested in the life of the Sun. Each and every second the Sun surrenders four million tons of energy in the form of light. She is expending all that she has, extending all that she is, in order for life to exist. As a result, Earth and all her inhabitants flourish in her radiance.

It seems that the giving over of life on behalf of ever-expanding creativity is integral to life itself. The massive star that was mother to our Sun met with fiery death, her form completely annihilated by the explosive force of the blast. And yet she exists in each of us, in the cells of our bodies that are composed her dust. Consciously or not, we carry her within us as surely as we carry the DNA of our biological parents. We are the children for whom she sacrificed all.

Of course we know that stars have no consciousness, no freedom to choose whether or not they will die, no matter how much creativity has sprung from their demise. But death is part of the process—every process, and that is the point. We have no choice about whether or not we die, just as we have no choice about whether or not we were born. What is crucial—what the Paschal Mystery teaches us—is that we can choose not to flee from death, but to meet it with grace.

In Matthew's gospel we see Jesus modeling a free and conscious response as the threat of death encroaches upon his life. The account of his final hours begins, "When Jesus had finished saying all these things, he said to his disciples, 'You know that after two days the Passover is coming, and the Son of Man will be handed over to be crucified'" (Mt 26:1–2). Aware that death is imminent—and what a cruel death it will be—Jesus presses on toward Jerusalem, choosing to participate rather than run for cover. The humanity of Jesus can become obscured as we enter the story ourselves. It is easy to allow Jesus to become a divine figure who is somehow separate from the suffering or, conversely, a supreme victim who remains passive and disengaged. But what is significant in this experience is precisely Jesus' humanity—his full flesh-and-blood humanity refusing to give over a single ounce of freedom and consciousness.

We usually focus on Jesus' death on the cross, naming that experience as his dying that saves us. What we sometimes forget to consider is that the cross was the final moment in a series of dyings that Jesus suffered as his human life came to a close. There was the dying of having to turn over his life's work to others—disciples who did not understand, who did not appear prepared for the undertaking. There was the dying to relationships. Mary and Martha and Lazarus had been friends of Jesus, and on his last visit Martha served a meal while Mary anointed his feet with aromatic nard. These friendships had been a source of support and love for Jesus, and like any human being he would have been greatly distressed to say goodbye.

Jesus suffered the dying of knowing that he was about to be betrayed by Judas, one of the disciples he himself had chosen. He went through the death of being rejected by one of his closest associates and trusted followers as Peter stood at the charcoal fire, within hearing distance, swearing to those around him, "I do not know the man!" Jesus went through the death of seeing all his men flee in fear, even after he had repeated to them time and again, "Do not be afraid." As he was dragged before both religious and political officials, Jesus experienced the death of having his message twisted by hate-filled accusers and blind judges. He experienced a dying as he watched the horrified face of his mother whose heart throbbed for her son and who would have gladly borne the lashes and the nails for him. There was dying that occurred as Jesus went to his death not knowing what was to become of his dream of the reign of God on Earth. There was even the dying of being cut off from Love itself as he cried out, "My God, my God, why have you abandoned me?"

Jesus met all these deaths with the awareness that is the fruit of the contemplative way in which he lived. As the narrative unfolds there is no report that he resisted by word or deed in any way, but appeared to meet each death with acceptance and without judgment. Perhaps all these dyings were met with such grace because throughout his life Jesus had practiced dying over and over and over again—those little deaths that we so much want to avoid. From the moment we encounter the adult Jesus—the one who is baptized and then goes to the desert to pray—we see one who rejects all that is not lifegiving, one who dies to self time and time again in order to serve the Mystery at the center of his life. Each of the little dyings releases the Spirit, and a lifetime of such choices manifests as a spiritual power that cannot be defeated by death, not even death on a cross.

In an earlier chapter we reflected upon Jesus as the one who signaled an advance in human evolution. He was able to capture the light of the Holy One in a way that transforms us all. His openness to the Light allowed us to encounter God's radiance in a new manner. His fidelity to the Light shows us how to live and die so that our lives are transformed here, now. He engaged death with every bit of consciousness and freedom that were his, and what we all discovered as a result is that death—while inevitable, while altering our dreams and causing us to let go of everything—does not have the final word. There is always—always—resurrection. And what is resurrection for us, in the context of the new universe story? It is a transformation in consciousness, an experience of transcendence in which we live out of the connectedness that is our truth. As we continue to evolve in consciousness, continue to emerge as more and more capable lovers, we share in the resurrection of Christ. We not only walk in the Light, we become light for others. Even little resurrections that come after choosing to die to fear and egocentricity release the Spirit. When we engage in a lifetime of death and resurrections as Jesus did, we become ever more empowered to do the work God asks us to do.

Life and death are a single mystery. That is what the Paschal Mystery teaches us. Death is inevitable—but so is resurrection. We can be sure that dyings will intrude upon our lives, and we may have some choice about how we can respond to their coming. We can be awake and watchful for the resurrections as well, for the creative ways that new life streams into our lives even in the midst of death. Like supernova explosions that shatter every recognizable fragment of life, we are capable of transcendence, capable of never allowing death to have the final say.

RADICAL AMAZEMENT

I think that one of the experiences that allows me to speak with conviction about dying is the experience of my father's death. He died of lung cancer, the result of contact with radiation at the government facility where he worked. He had known of his exposure and was aware of its consequences, a fate shared by many of the men who worked there, including two of my uncles. Cancer deaths were common among them all. When I found out about the source of his illness, I was deeply angry and resentful at the carelessness and disregard for human life that seemed to be the standard operating procedure at the plant. As I shared my anger with my husband, he told me about a conversation he and my dad had not long after his diagnosis. In the face of death, rather than seethe with resentment, my dad had been able to forgive those who had placed him in danger and then lied about it. Learning of my father's forgiveness softened my anger and began my journey to forgive.

In the last weeks of his life Dad, my mother, my sister, and I had the privilege of entering into his death, not with resistance or bitterness, but with trust in the Mystery we knew was at the heart of the process. It was incredibly painful—for him, physically of course, but for all of us as we let go of dreams and a man whom we loved so, and whom we knew loved us as deeply as a husband or father could. During his final days he pointed to the beings of light that he saw in the room, and whenever he saw them the look on his face was of sheer delight. "I feel so free," he said to us, "I feel so free I could fly." When he breathed his last I had the profound sense that he was now fully in the presence of the Light, that he was one with it now, and in the radiance of that light we would know his presence forever. We grieved, and our hearts filled with sorrow, but there were signs of life everywhere. Six weeks after Dad's death Ashley was born, the second great-grandchild of my parents, and with her came the reminder that life is precious, that resurrection comes even in the midst of the pain of death.

Death is woven into the fabric of the universe, an integral part of the Mystery of life itself. When we step back and view the big picture, we know that death and life are inseparable. Both are linked to the creativity and release of Spirit that moves life along. What is not inevitable is being ruled by fear. The experience of fear will surely come, but it does not have to rule us, does not have to control our response or drain our power. When we embrace our freedom and live in a contemplative way that expands our consciousness, we can respond to fear in a way that allows us, like Jesus, to engage all our deaths and dyings in a way that is radically amazing. In every moment of death there is a release of the Spirit, and in every movement of Spirit there is resurrection—and life.

WE COME TO CONTEMPLATION

* *Imagine that you are a witness to the supernova that gave birth to our Sun. See the giant burning star grow redder and redder and then die as it explodes into billions of pieces, rippling space and spewing cosmic debris in all directions. Acknowledge the loss that this death has brought about. . . . Envision the debris from the supernova interact with a great cloud of hydrogen, and then see bits and pieces pulled together by gravity and take the shape of a new star. Watch as the Sun cools and begins a regular rotation that holds planets in orbit. . . . Identify the planets that you know: Mercury, Venus, Earth, Mars, Jupiter, Saturn, Uranus, Neptune, Pluto. . . . Bring your attention back to Earth. See how her outer crust hardens, the seas gather, and life begins to emerge. . . . Become aware of how closely connected our life is to the Sun. . . . What new perceptions about life and death are emerging in you? What is it you want to say to the Creator?*

* *What are the "supernova experiences" of your own life? Where have you encountered the life-death-life-death movement that is a painful yet necessary part of the universe?*

* *What in this chapter do you find challenging? What do you find radically amazing?*

* *Sit quietly for a moment. What does the Spirit want you to see? How do you respond?*

WE PRAY

God of all creation, we acknowledge that death is a mystery that we often resist, failing to accept that it is indeed part of life. We find dying difficult, partly because we are afraid, partly because we treasure the life you have given us. Help us to accept the reality of death with grace, just as Jesus did. Help us to enter into our physical death and all the dyings that precede it with as much consciousness and freedom as we can. May we not judge ourselves or others harshly as we engage in death, but trust that your presence will guide us through the mystery and into your heart of love. Amen.

CHAPTER ELEVEN

MYSTERY: DARK ENERGY AND DARK MATTER

I form the light and create the darkness.

* ISAIAH 45:7

The apparent acceleration of the expansion of the universe is attributed to a dark energy residing in space itself.

* STEVEN WEINBERG

All we can know for sure is that a mystery courses through us, seeking its own fullest incarnation, and that whenever we serve the mystery within, we experience a linkage to the mystery outside. When we stand in more conscious relationship to this mystery, we are more deeply alive.

* JAMES HOLLIS

One of the most amazing discoveries of the twentieth century is that the universe is expanding. It is not that the universe is expanding in space and time, but rather spacetime itself is expanding. As noted earlier, Albert Einstein and his contemporaries were quite comfortable with a universe that was static and contained. Einstein's own theory of relativity, much to his dismay, implied that the universe must be either expanding or contracting, but could not exist in a steady state. In 1929 Edwin Hubble provided proof of expansion by measuring a phenomenon known as redshift,[1] for all intents and purposes silencing those who wanted to believe the contrary.

It was generally assumed that the expansion of the universe occurred at a constant rate, but during the 1990s astronomers began to notice that galaxies at the farthest edges of the universe were dimmer than predicted, indicating that they were farther away than previous calculations indicated. The new data suggested that not only was the universe expanding, but that the rate of expansion was increasing. In 1998 two separate teams of scientists, the Supernova Cosmology Project at Berkeley Lab and the High Redshift Supernova Team at Harvard, confirmed that this was the case. Not only is the universe expanding in all directions, but the rate of the expansion is accelerating.

The term used to describe the force that causes the universe to expand has been named "dark energy." Dark energy is an antigravitational force that cannot be seen, but its presence is inferred from the effects it produces. According to Sean Carroll of Fermi Lab in Chicago, dark energy appears to be spread rather uniformly throughout space and maintains a constant density as the universe expands.[2] In other words, dark energy does not seem to change, but remains constant throughout both space and time. Some scientists suggest that its existence puts to rest a theory that depicts a universe that will eventually implode upon itself if it remains under the influence gravity alone.

Recent discoveries affirm that dark energy is found in all reaches of the universe, from its farthest edges some twelve billion light years away, to the local group of galaxies that contains the Milky Way. Fabio Governato of the University of Washington says that the data that has been gathered gives a picture of a universe that is a sea of dark energy with billions of galaxies emerging like islands.[3]

Perhaps the most amazing detail of the theory of dark energy is that it makes up about seventy percent of the universe! Over two-thirds of the cosmos exists in a form we cannot see or touch, yet it exerts a force which cannot be denied. As an anti-gravitational force, it contributes in some mysterious way to ongoing creation, to sustaining life and providing the necessary components for continued development.

But dark energy is not the only dark mystery of the universe. As observation technology advances, scientists have been able to gather data from galaxies from the deepest regions of the known universe. Those who study these galaxies began to notice something interesting about their rotation. Some galaxies, it was discovered, rotate at a speed of 60,000 miles per second. The obvious question to ask is, "What keeps them from flinging apart?" The answer: "dark matter."

Dark matter cannot be seen but is known through the gravitational pull it exerts. "Dark" means that the matter is neutral, carrying no electrical charge that enables it to interact with light.[4] It appears that about twenty-five percent of the universe is comprised of dark matter. Whereas dark energy is anti-gravitational and is spread uniformly throughout space, dark matter is dense and contributes to the gravitational field of a galaxy or cluster of galaxies.

The attempt to understand dark energy and dark matter is one of the most significant puzzles that current researchers are trying to solve. Saul Perlmutter, who heads up the Supernova Cosmology Project bluntly puts it, "The universe is made mostly of dark matter and dark energy, and we don't know what either of them is."[5] Most of the universe, in other words, is mystery. Most of it—as much as ninety-five percent—cannot been seen or touched, yet all life exists because dark matter and dark energy are there, bringing things together in wholeness or stretching them apart in ever-expanding creativity.

Many of us are uncomfortable with mystery. Mystery has to do with befriending darkness. Mystery has to do with not knowing, with unknowing, with living in unprejudiced awareness in the present moment with nothing to hold us save our trust in what is unseen. Mystery requires that we negotiate the darkness, aware of the force that holds us together, sensitive to the pull that sends us forth. Often we resist mystery of any sort, perceiving the unknown and uncertain as threats to be eliminated rather than invitations to deeper truth. Safety needs can often take precedence over spiritual needs, and orthodoxy and orthopraxis can become hiding places from which to escape uncomfortable questions and nagging doubts. We do not like the darkness! Even as we profess that God is Mystery, we resist surrendering to the intangible that we label as frightening or unreal.

Mystery calls us not only to lay down our lives, but to lay down our agendas that interfere with our call. Mystery invites us to live with wisdom, to know when to stand firm and when to take flying leaps. Mystery asks us to live in the unknown with faith and to live in the uncertain with hope, trusting in the revelation of a deeper knowing and certainty that manifests as Love.

We should be uncomfortable with Mystery. After all, it will not allow us to escape anything that is less than lifegiving. It will not allow us to play games or play small, and it will expose the places where we hide with fright. Mystery asks us to surrender our egos and fears, our prejudices and agendas, and then free fall into Love. It will take every ounce of humanity we have and every ounce of grace God gives to live in fidelity to Mystery's beckon, and yet for this we have been born. For Mystery and Wisdom and Love we have been born. For awareness that allows creation to be conscious of itself we have been born. For life that incarnates the Spirit, we have been born. Mystery is a must, our participation required.

The Holy One is Mystery, Mystery is the Holy One. How often it is said that if we think we know, what we know is certainly not God, who is beyond all that we grasp with our minds. God is beyond rules and regulations that restrict freedom, beyond dogmas and doctrine that instill fear. Just as the vast majority of the universe is beyond our capacity to see or touch directly, so is the divine. But like dark energy and dark matter, we can see signs of God's presence all over the cosmos. Just as dark matter works as an unseen gravitational force to hold galaxies together, so Mystery's presence draws us to the heart of a Holy Darkness. It gifts us with intelligence and personality, with relationships and work, with purpose and meaning. Mystery understands who we are and all that we can be. Holy Darkness draws us together, making us whole, and when we operate out of a place of wholeness we become luminous beings that radiate the Spirit's presence and light up the lives of others just as surely as any star.

But the work of Mystery manifests in another way as well. Just as dark energy is an antigravitational force that expands the universe, so will the Holy One cause us to move out of our place of comfort and into space that is uncharted and unknown. In the process we are part of the evolution of the cosmos. Mystery not only desires but demands that we participate in life. Our very being is a being-in-connectedness. What we must decide is how we will do our part—whether we will be fear-filled resisters or love-based risktakers. As we ride the dark energy into unknown territory, as we say yes to Mystery that is beyond grasp but not beyond grace, we incarnate the Spirit and fulfill our role as co-creators of a universe that is expression of pure Love.

In his poem "The Dark Night," John of the Cross offers insights about living in Mystery.

130   RADICAL AMAZEMENT

One dark night,
fired with love's urgent longings
—ah, the sheer grace!—
I went out unseen,
my house being now all stilled.[6]

John's poem is about the soul's union with God, and the dark night of the soul refers to the growth and development that transpires in darkness—in the unseen and hidden night places of our lives. All of us are fired with love's urgent longings because we are part of the original flame that brought all life into being. Our urgency comes from the awareness of the conditions that lie around us and within us. The mystic's poem is not only for the specially-called, the select, the ones we label "holy." "The Dark Night" is about us, an expression of the secret process that forges our identity.

John steps out, propelled into darkness at the proper time, "my house being now all stilled." John's walk in darkness begins not in a tempest of chaos, but in stillness. He comes out of a place of contemplative silence, a place of peace in which he knows as fully as he possibly can the dimension of God's love and fidelity. He does not go out frantic with fear or full of doubts. He strides confidently into the Mystery, not knowing but not resisting what lies ahead. He trusts in the darkness, trusts in the presence of the Beloved to lead him in sheer grace.

O guiding night!
O night more lovely than the dawn!
O night that has united
the Lover with his beloved,
transforming the beloved in her Lover.

In the stillness of the night, in our convergence with Holy Darkness, we are united with the Mystery that lies in the cavern of our heart. "I abandoned and forgot myself," the poet tells us. It is in the abandonment and forgetting he has been filled, remembering in truth who he is.

On Holy Thursday our worshiping congregation sits in hushed silence. With lights extinguished, we begin to sing:

Holy Darkness, blessed night,
heaven's answer hidden from our sight.
As we await You, O God of silence,
We embrace your holy night.[7]

Life is a Mystery shrouded in darkness. But the darkness is fecund, a place of possibility and power. What is radically amazing is that we are invited into the darkness, into the heart of creation and creativity, invited to participate—unsure, but confident in the invitation and the One who extends it. May we, like John of the Cross and Jesus, embrace the darkness and become radiant with life.

WE COME TO CONTEMPLATION

* *Imagine that you have traveled to the farthest reaches of the universe, some twelve billion light years out into space. . . . Notice as you travel that there are clusters of galaxies that are grouped in space, pulled together by the force of gravity. Observe how they spin so fast that if it were not for the force of dark matter, they would be flung apart. . . . Watch as they move, connected by an unseen presence that binds them. . . . And now notice that although these galaxies are connected, the universe is expanding—space and time themselves are expanding. . . . Recognize that the expansion is being propelled by dark energy that you cannot see but whose effect is everywhere. . . .*

* *And now take a look at your own life. Recognize times that the Mystery pulled you together, brought you into greater wholeness and cohesiveness, in and through the darkness. See the times God surprised you with insights or movements that you know were part of the deep Mystery at the heart of the universe that invests itself in details. See the times that you have taken*

risks, propelled by the Spirit to participate in the dark Mystery in a way you could never have imagined.

* *What in this chapter do you find challenging? What do you find radically amazing?*

* *Sit quietly for a moment. What does the Spirit want you to see? How do you respond?*

WE PRAY

Holy Darkness, God of Mystery that creates in and through us, help us not to resist the darkness. Help us to trust in you precisely in those moments when we are confused or uncertain or do not understand. Heal us of our trepidation in the face of the unknown, and help us to yield to the creative process that at this very moment is at work in the inner darkness, in the unseen, secret places that only you know. Like dark energy, may your Spirit expand my being so that I become more of who you desire me to be—free, capable of loving and being loved. Amen.

CHAPTER TWELVE

THAT ALL MAY BE ONE: LIVING IN RADICAL AMAZEMENT

Within each of us lies a power greater than anything we will ever have to face. Sometimes it reveals itself as a great force—at other times, as a deep, unfolding stillness. For centuries it's been described as a burning fire. Yet this power never imposes itself. Each heart decides to respond to its presence—and that decision sets the course for everything that follows.

＊PAULA D'ARCY

There are lots of ways of being fearless. I highly recommend it.

＊CHRISTOPHER REEVE

Someday after mastering winds, waves, tides and gravity, we shall harness the energies of love, and then, for the second time in the history of the world, man will discover fire.

＊TEILHARD DE CHARDIN

A flourishing humanity on a thriving Earth in an evolving universe, all together filled with the glory of God—such is the theological vision and praxis we are being called to in this critical age of Earth's distress.

＊ELIZABETH A. JOHNSON, CSJ

We return to the place where we began, remembering that Thomas Aquinas said that a mistake in our understanding of creation will necessarily cause a mistake in our understanding of the divine. The guiding motif of this book has come from Abraham Heschel's assertion that our response to any encounter with the divine should evoke from us a response of radical amazement. This radical amazement, as Dorothee Soelle suggests, tears apart the veil of triviality, revealing the significance and magnificence at the heart of even the most ordinary aspects of life. Soelle also claims that radical amazement is the starting point for contemplation, both a practice and a way of living characterized by attentiveness to what is and responsiveness to what seeks to be.

In each chapter of *Radical Amazement* we have held one small piece of the new universe story, examined it with attention and become aware of ways that it may reveal just a little bit more of the divine to us and in us. Our pursuit has not been technical data or scientific facts, but the Holy One. Hopefully there have been moments of radical amazement for you, followed by reflective silence that has deepened the sense of the Holy in your heart and in your life. Perhaps you have begun to look a little more attentively to your own life and found places of radical amazement all around.

We have used the discoveries of modern science to bring us back to nature—to the raw beauty of creation and to human nature. Our faith tradition has always used images of creation—from mountains and seas to seeds and light—as metaphors that help us wrap words around the ineffable. This book uses new images and metaphors to expand and deepen our relationship to the Holy One, to one another, and to all creation in a time in which all these relationships appear to need a great deal of support.

In Jesus' final discourse he found it necessary to speak to his disciples about unity, "That they may be one, as we are one" (Jn 17:11, 22). To add emphasis Jesus prays these words twice. Indeed, they are the heart of his mission and message. Jesus' radical unity with the Holy One defined his life, and his prayer indicates that he wants that same radical unity to define those who follow. The reflections in *Radical Amazement* have offered images that enable us to see the connectedness and live in a unity (not uniformity) that allows us to celebrate the beauty of diversity as we commune with one another and function as a created and creative whole.

Significant questions remain: How is it possible to live as a created and creative whole? What will help us inhabit this vision and make it concrete? Again we turn to science and the new universe story for a response. According to some scientists learning that the universe is expanding was the greatest scientific discovery of the twentieth century, and the implications of that discovery will engage explorers well into the new millennium. It seems that this image—of the universe flinging itself out in beauty and splendor and infinite creativity—provides a fitting response to these questions.

We live in a time that calls us, just like the universe, to expand—to fling ourselves out into life with creativity and zest. We are invited to be sparked by the primordial fire from which we came. We ourselves are to become flames of love that burn away fear and bring the warmth of compassion to all creation. Our expansion will require that we embrace our gifts and capacities for co-creativity as well as reject anything that works against what calls us to life. The various kinds of expansion and embrace that will enable us to live in unity now need to be considered.

The new universe story asks us to *expand our image of God*. Over the last few centuries we have all but forgotten the God who is incomprehensible Holy Mystery. Instead the God who has been front and center often has the characteristics of a divine mechanic working from outside the created world, tinkering when necessary and performing scheduled maintenance, but too often removed from a life that froths and foams with vitality. Even when we call God "Father," an image intended to convey love, we may perceive God at a distance, enthroned in heaven rather than rooted in Earth. I recently heard the story of a group of children from mixed ethnic backgrounds who were interrupted from their play and asked the question, "Where is God?" The Catholic children pointed up into the sky. The Hindu children pointed to their hearts. While it is significant to retain the notion of God's transcendence, we must not do so at the expense of God's imminence. God's presence is woven throughout all creation, manifest in all that is. Our image of God needs to be at least as big as life, as expansive as the universe.

Expanding our image of God invites us to *embrace a God of radically amazing Mystery*, One who continuously gives expression from within creation, One who emerges as new life in every nook and cranny of creation. We are asked to reject the anthropocentric projection that makes God into our image. It is human beings who are made in the Creator's image and likeness—a God characterized by freedom and compassion, fidelity and service, justice and mercy. When we reverse the movement, projecting our viewpoint onto God, these characteristics often fail to appear. To embrace a God of mystery means that we live open to surprise, ready to respond in radically new ways to the One who never ceases to amaze. Embracing a God of mystery means that we live in relative comfort with the unknown, grounded in the assurance that the divine is ever present, ever at work in wondrously creative ways.

The new universe story also invites us to *expand our hearts to include all creation*. Throughout this book, from the theory of holons to the possibility of morphogenic fields, we have looked at the ways in which all of life is connected. Indeed connectedness is fundamental to our reality. No matter which sphere of life we observe, from the physical to the spiritual, we are connected to others. We cannot separate from this truth, no matter how hard we may want to try. Albert Einstein called the notion of separation an "optical delusion," yet for several centuries now we have lived deluded, and at great price. So many of the social and ecological problems that confront us today stem from our delusion that we are separate from, better, or more significant than, other members of creation—from other groups of people we encounter to the air we breathe. Our lack of openness to all may very well mean our demise.

If we are to expand our hearts to include all creation we need to *embrace our capacity for communion*. Meister Eckhart said, "Relation is the essence of everything that is."[1] How true. Relationship is something that all life requires, even inorganic life. Our vitality depends upon the connections we establish and the communion we share. Of course we are made for agency, of course we are asked to develop our gifts and use them well. But our gifts, as lifegiving as their full expression can be for us, are foremost for others. This means that we must reject a perception of separateness and exclusivity that keeps others at bay. And it means we reject any temptation that keeps us from self-communion, from tending to the Holy who dwells within. What nourishes any of us, more than bread itself, is a relationship in which we discover simultaneously who we are as we discover who the other is. Communion that honors the other, that reverences the Holy One in the other and in the self—this is what we embrace. Connectedness is primary. Communion is essential.

The new universe story invites us to *expand our commitment to emergence*, to participate in the divine unfolding around us and within us as fully as possible. The process of evolution reveals life as unfinished. All life is in flux, all life is groaning toward fuller expression and greater consciousness. Being committed to emergence means that we engage the world around us, with the Spirit finding us approachable and accessible partners in co-creation. We look for new life and nurture it where we can. We challenge ourselves to be open and grow from *Homo sapiens* into *Homo universalis*, universal humans who integrate mind, body, and spirit in a way that is lifegiving to all.

Emergence implies the unfolding of that which has never yet existed. It requires great trust in the Spirit, for there are no maps to guide us as we venture into the unknown. Like explorers in virgin territory, we make our way carefully, trusting in the life skills we have acquired and the intuition that urges us on. We are ready to change directions when necessary and take risks when required. Safety and security take back seats to creativity and courage. Uncertain about where the journey will take us, but sure that we must press on, we participate in the emergence of a new era of humankind, one that embraces all of life. An expanded commitment to emergence asks us to be open and patient with the new that is evolving within us as well, acknowledging that Mystery speaks as powerfully in darkness as in light.

Expanding our commitment to emergence requires that we *embrace our capacity for self-transcendence.* As holons with consciousness, we are able to move to new levels of awareness. We are capable of negotiating the tension between self-preservation and self-adaptation in a way that affirms both our agency and our connectedness to all that is. The capacity for self-transcendence means that we are able to evolve in complexity. Our cognitive, moral, and ethical awareness can develop greater depth as we engage life and grow in our ability to know what is valuable and what is peripheral. We have the ability to question and to dream, to expand our vision and live out of other-centered concern. Our capacity for self-transcendence asserts that healing is possible, that we are more than our wounds and our scars. As we engage this capacity, we grow in freedom. As we embrace the process that allows us to transcend, fear begins to fall away as we recognize our value as part of the whole that is continuing to emerge.

Self-transcendence sets us free—free to respond lovingly and creatively as we enter into the flow of emergence.

The new universe story also invites us to *expand our capacity for empowerment*. Marianne Williamson once wrote that our greatest fear is not that we are inadequate. Our greatest fear is that we are powerful beyond measure. I agree with her assessment. In all the universe, human beings are the ones empowered enough to ask "What if?" and "Why not?" Human beings are made in the image and likeness of God. Human beings are the universe conscious of itself. Human beings are the ones who have the heart to feel compassion and the wisdom to do justice. Human beings alone have heard the call to live in relationship with the divine, to live in fidelity to the truth and to live in service of love. We have been given power and the ability to recognize and own it.

The life and death of Jesus shows us that we have been given everything we need to become receptors of the Light that is Life itself. Jesus' call for us to become light to the world is not an empty invitation but a potent promise. The call challenges us to move out of our complacency and demands that we live in the awareness of our connection to all. Each moment of our lives we are linked to others, a holon within other holons, interdependent yet free. Every choice we make affects a morphogenic field that sets the possibilities to come. It is undeniable: We are potent and powerful!

To fully engage the power that is ours, we must *embrace our capacity for agency*. The capacity that we have to connect and remain in relationship with others is balanced by our capacity to stand alone, refusing to give control over to what is not lifegiving, standing firm while doing the work that needs to be done. Each of us has gifts that are needed for the whole. "Remove from your midst oppression, false accusation and malicious speech," we read in Isaiah (58:9). By reaching out to those around us, seeing where justice and mercy are absent, we exercise the capacity for agency. When we counter false accusations that someone—including ourselves—is not essential to the whole, we

RADICAL AMAZEMENT

become agents of truth. When we reject misguided speech that holds that economic development or military solutions are the answers to all our problems, we become agents of light.

Agency does not mean that we do it all, but simply that we do our part. In the second letter of Paul to Timothy, the writer tells the disciple to fan into flame the gift of God that he received, a gift that is not characterized by fear, but by power, love, and self control. (2 Tm 1:6–7) We truly are empowered by the Spirit of the Holy One. This power, forged by the Spirit and tempered by self-control, enables us to give love away. That is what our self-transcendence and freedom are about. Powerful beyond measure, we can do with love what we will.

The new universe story invites us to *expand our contemplative awareness.* Expanding our image of God, embracing all of creation, being committed to the process of emergence, and living empowered—all our efforts must flow out of contemplation or they will be doomed to fail. Why? Because as wonderful a species as we are, we are nothing without grace. God calls us to be co-creators not because we have earned the right, but because we are able to receive the gift. Our ability to be aware, to be attentive, to see and to hear, to feel and to intuit—all come more fully alive when they flow out of a quiet, receptive heart. And so our participation in ongoing creation is predicated on our commitment to silence—to cultivating a listening heart and a peaceful spirit.

Living in contemplative awareness invites us to *embrace our capacity for awe.* The radically amazing surrounds us. In each and every moment the Holy One seeks to reveal a truth. Something sacred is always afoot. But we often miss the moment. With many things clamoring for our attention we have grown so weary of looking that we soon tune out everything. Overstimulated, we have grown too numb to notice beauty, too worn out to engage in wonder. Like the primordial cells unable to capture the rays of the Sun, we languish in darkness while light bounces all around.

At the heart of contemplative living is a reengagement with awe. Abraham Heschel says that awe is "a way of being in rapport with the mystery of all reality."

> Awe enables us to perceive in the world intimations of the divine, to sense in small things the beginning of infinite significance, to sense the ultimate in the common and the simple; to feel in the rush of the passing the stillness of the eternal.[2]

Awe begins with at least a momentary cessation of activity in order to behold what is present, what endeavors to evoke a response, to elicit a signal that we have seen the sacred. Awe enables us to recognize the connectedness of all things and live in the humility that flows from knowing how connected we are. As we grow in the capacity to recognize the holy in each moment, we create a morphogenic field that enhances all of life, not merely our own. Awe draws us beyond ourselves, opens our hearts, and allows us to tumble head over heals into Mystery that seeks ever-greater revelation and communion. Fidelity to awe—the commitment to be attentive to the many ways life is emergent around us—enables us to live contemplative lives, and living contemplative lives keeps us open, keeps us evolving, keeps us connected to our universe and to our God.

A Life of Radical Amazement

What will life look like for us if we expand our image of God, envision life as including all creation, commit ourselves to emergence and empowerment, and live in contemplative attention? Can you imagine how radically amazing life will be? Do you think it is possible? Are you willing to risk your life for it?

A life of radical amazement suggests that we live at a slower pace. We spend time doing "nothing." Learning to sit still, we become aware of all that surrounds us. We pay attention to singing birds and playing children. We notice clouds and the signs of changing seasons. We become friends with trees and neighbors. We begin to relax, and tension starts to melt away as we tune in to the energies of Earth. Slowing down, we notice interior movements. Feelings surface, and sometimes they are not pleasant ones, but as we listen to their message and bring the healing energy of the Spirit through Earth, we hear their message and acknowledge the benefit of being in touch with our truth, even if it is painful.

Living with increased awareness and sensitivity we are not so likely to get sucked into the black hole experiences that drain our energy and set us on a collision course with disaster. Behaviors that lead to codependency and addictions or in other ways restrict our freedom are not as likely to take hold when we are filled with wonder and awe. When the inevitable black hole experiences do come, we can meet them with more courage and strength. In those times when we are overwhelmed by the darkness, the wisdom in our hearts, accessible through contemplative attention, will usually assure us that even in the darkness there is the possibility of a breakthrough of light. And then, in contemplative awareness, we have the ability to wait for the coming of the light with grace.

Living in radical amazement means that we also recognize our unique time and place in history. We are the first generation to know empirically that the universe came about through a powerful emergent event 13.7 billion years ago. It is only in our lifetime that images of billions of galaxies, each with billions of stars, the remnants of the original flaring forth, have been seen. The Hubble Space Telescope images show us with incredible clarity just how exquisite the universe is. We are the first to see Earth from outside its atmosphere, a delicate blue marble suspended in space with no artificial boundaries, only a unified biosphere of indescribable beauty.

As we live a life of radical amazement, we are sensitive to the process of emergence and the possibility of evolving from *Homo sapiens* to *Homo universalis*, universal humans who are integrated body, mind, and spirit. Our understanding of Jesus as the one who embodies the evolutionary advance that enables to us become receptors of the Light of the Creator helps us to become clear about who we are. We know that we are beings of light and energy who form morphogenic fields that create possibilities for others. We are holons nested in one another, holding one another, managing the tension between agency and communion, transcendence and dissolution. We acknowledge that death is a painful though necessary part of life and that life itself unfolds in and through Mystery that is incomprehensible yet present. We know that unity is the underlying truth and that wholeness is the universal drive.

All of our knowledge leads us to greater consciousness. Our knowing what we know is an act of self-transcendence, and our acting upon what we have learned will lead to greater consciousness still. We must now become intentional about responding to the challenges of living in an emerging universe. We must accept the power and grace that is there, that has been there, for a very long time. This moment in time is for us like the moment the first chlorophyll cell learned to receive a ray of light from the Sun and became a source of nourishment and life for all living beings. This is our moment. Let us live connected and in love, so that generations to come will look at us and say, "They were the first generation to really get it. They were the first universal humans, the first to take in the universe and hear its story and know their part in it. They were the first to make choices rooted in the conviction that all life is connected. They were the first to receive the Light in a way that allowed the entire species to escape the pull of dissolution and disconnectedness and transcend to a new level of vitality and freedom that changed the whole of creation. They were radically amazing!"

WE COME TO CONTEMPLATION

* *Imagine that you are the universe. See that everything within you is connected, every star and galaxy, every black hole, all the dark energy and dark matter. . . . Look at the solar system within you, see planet Earth and all her beauty, from her mountains and seas and rivers and forests to each creature that makes their home here. . . . Absorb the reality that everything is connected, that in this vast Mystery all is one— and this vast Mystery resides in you. . . . Allow yourself to be swept up in awe. . . .*

* *How is your image of God expanding? How does an expanded image change how you relate to God? Where is the Spirit inviting you to grow?*

* *How are you expanding your heart to include all creation? What are concrete examples of your expansion? Where is the Spirit inviting you to grow?*

* *How committed are you to the process of emergence? How have you witnessed emergence in your own life? How will you prepare yourself for ongoing emergence? Where is the Spirit inviting you to grow?*

* *How is the Spirit inviting you to expand your power? What specific tasks are you empowered to do? How will you tend to your power so that you may grow in the ability to use it wisely? Where is the Spirit inviting you to grow?*

* *What have you learned as you have taken in the insights from Radical Amazement? Bring any new awareness within you to the Holy One, hold it there, and listen to what God says to you.*

WE PRAY

Holy One, in every moment we live in your expansive love and your tender embrace. All around us we behold your presence. All around us and within us life emerges, fresh and new, vital, sparked with zest. May we continue to expand our lives and our living, accepting the challenge and the responsibility to be co-creators with you. May we embrace all the gifts you have given and use them well, in love and for love. May we use our gifts with wisdom and with courage, celebrating the connectedness and seeing the expressions of your love everywhere. May our actions and attitudes generate unity and our love bring us to communion. May we live in such a way that generations to come will look at us in you and say, "Radically amazing!" Amen.

GLOSSARY

Big Bang: The generally-accepted theory that accounts for the origin of the universe 13.7 billion years ago. The universe came about as a result of an explosive burst of energy from a particle smaller than an atom.

Black Hole: A region of space, at the center of which is a singularity so dense that nothing that has passed a border called the event horizon can escape, not even light.

Causative Formation: Rupert Sheldrake's theory of morphogenic fields; postulates that morphogenic fields are unseen forces that preserve the form of self-organizing systems.

Cosmic Microwave Background Radiation (CMBR): Discovered in 1964, radiation at the microwave range that is evidence of the Big Bang.

Cosmology: The story that flows out of the study of the origin and development of the universe, including purpose and meaning.

Cosmos: The entire universe.

Dark Energy: The antigravitational force that causes the rate of expansion of the universe to accelerate. Seventy percent of the universe is dark energy.

Dark Matter: The gravitational force that holds galaxies together. Dark matter cannot be seen but its effects can be measured. Twenty-five percent of the universe is dark matter.

Emergence: The process of becoming more conscious and more complex in the course of evolution. All holons emerge.

Escape Speed: The velocity that is required for an object to escape gravitational pull and continue into space forever. Earth's escape speed is about seven miles per second.

Event Horizon: In the region around a black hole, the point past which escape is impossible.

Evolution: The process of development into different and generally more complex and fit forms.

Gaia Hypothesis: Postulated by James Lovelock and Lynn Margulis in the 1960s, this hypothesis asserts that Earth is a single, living system or biosphere.

Hawking Radiation: The theoretical possibility of escape from a black hole as a pair of particles split, one entering the black hole and the other splitting off; the action of the particles in this way may cause the black hole to lose mass and eventually collapse.

Holon: A whole/part. All beings are holons, distinct modes of being (wholes) that are part of a more complex mode of being. The four characteristics of holons are self-preservation, self-adaptation, self-transcendence, and self-dissolution.

Morphogenic Field: In the human species, morphogenic fields set up habits of thought, activity, and speech. Morphogenic fields (from the Greek, *morphe*, "form") are not energy fields, but help manage the energy of a system by carrying information that maintains its wholeness.

Photon: A particle of light.

Prokaryotes: Primitive unicellular organisms, the evolutionary ancestors of complex organisms.

Quanta: A discrete packet of energy, a photon.

Quasars: Acronym for "quasi-stellar radiators" which carry the energy of millions of stars; associated with black holes.

Redshift: While light moving toward us shifts to the blue end of the visible spectrum, light moving away shifts to the red end of the visible spectrum. By measuring redshift, physicists can determine the age of the universe and how fast it is expanding.

Self-organizing System: A form or structure that maintains itself from within.

Singularity: The super-dense particle at the core of a black hole.

Supernova: The death explosion of a star.

Special Relativity: Einstein's theory that revolutionized the scientific understanding of space, time, and energy.

RADICAL AMAZEMENT

NOTES

INTRODUCTION

1. Thomas Gilby, *St. Thomas Aquinas: Theological Texts* (Durham, England: Labyrinth Press, 1982), 76.
2. Terence Dickinson, ed., "Astronomy 2003" calendar (Buffalo New York: Firefly Books, 2003).
3. Abraham Joshua Heschel, *God in Search of Man* (New York: Farrar, Straus and Giroux, 1955), 117.
4. Ibid.
5. Ibid., 46.
6. Ibid.
7. Dorothee Soelle, *The Silent Cry* (Minneapolis: Fortress Press, 2001), 89.

CHAPTER ONE: COSMOLOGY

1. Brian Swimme, *The Hidden Heart of the Cosmos: Humanity and the New Story* (Maryknoll, NY: Orbis Books, 1996), 98.
2. My primary source for the discussion of the history of cosmology is "Einstein's Relativity and the Quantum Revolution," a series of lectures by Professor Richard Wolfson of Middlebury College (Chantilly, VA: The Teaching Company, 2000). Dr. Wolfson's lectures provide a basic overview of quantum physics for the non-scientist.
3. It was not until the Renaissance that the third component of the scientific method was added: systematic testing to prove or disprove an observation.

CHAPTER TWO: THE NEW COSMOLOGY

1. A major source for this chapter comes from lectures given by Professor Steven L. Goldman, Lehigh University, "Science in the Twentieth Century: A Social-Intellectual Survey" (Chantilly, VA: The Teaching Company) 2004.
2. This information was gathered from the following website: http://physicsweb.org/articles/world/17/5/7/1. Feature article, May 2004. (author not cited.)
3. Steven L. Goldman, *Science in the Twentieth Century: A Social-Intellectual Survey* (Chantilly, VA: The Teaching Company, 2004), 24.
4. Jeffrey G. Sobosan, *Romancing the Universe* (Grand Rapids, MI: William B. Eerdman's Publishing Company: 1999), 4.
5. Brian Swimme, *Hidden Heart*, 36.

CHAPTER THREE:
IN THE BEGINNING: THE UNIVERSE FLARES FORTH

1. There are a number of websites that offer visual representations of the Big Bang. Among my favorite is the one operated by PBS. Check out www.pbs.org/deepspace/ timeline and www.pbs.org/wnet/hawking.

2. Bede Griffiths, *A New Vision of Reality: Western Science, Eastern Mysticism and Christian Faith* (Springfield, IL: Templegate Publishers, 1989), 30.

3. Rosemary Ellen Guiley, *The Quotable Saint* (New York: Checkmark Books, 2002), 4. See also Louis J. Puhls, S.J., *The Spiritual Exercises of St. Ignatius* (Chicago: Loyola University Press, 1951), 102.

4. Ibid.

5. James Finley, *Christian Meditation: Experiencing the Presence of God* (New York: HarperSanFrancisco, 2004), 24.

6. Audio recordings include *Christian Meditation* by James Finley (Boulder, CO: Sounds True) and *The Contemplative Journey* by Thomas Keating (Boulder, CO: Sounds True).

CHAPTER FOUR: LET THERE BE LIGHT: THE BIG BANG

1. Steven L. Goldman, Lehigh University, "Science in the Twentieth Century: A Social-Intellectual Survey" (Chantilly, VA: The Teaching Company), 2004.

2. Gerald L. Schroeder, *The Science of God: The Convergence of Scientific and Biblical Wisdom* (New York: Broadway Books, 1997), 165.

3. Ibid.

4. Elizabeth A. Johnson, *Women, Earth, and Creator Spirit* (Mahwah, NY: Paulist Press, 1993), 48.

5. Peter Hodgson, *Winds of the Spirit: A Constructive Christian Theology* (Minneapolis: Fortress Press, 1994), 279 in Diarmuid O'Murchu, *Evolutionary Faith: Rediscovering God in Our Great Story* (Maryknoll, NY: Orbis Books, 2002), 48.

CHAPTER FIVE: ALL CREATION IS GROANING: THE PROCESS OF EVOLUTION

1. Teilhard de Chardin, *The Human Phenomenon,* Sarah Appleton-Weber, trans. and ed. (Portland, OR: Sussex Academic Press, 2003), 190.

2. Diarmuid O'Murchu, *Reclaiming Spirituality* (New York: Crossroad Publishing Company, 1997), 99.

3. David Quammen, "The Evidence for Evolution Is Overwhelming," *National Geographic,* 206, 5, (Nov 2004), 20. In this description of evolution I draw heavily on this article.

4. Ibid., 4.

5. Ibid.

6. John Paul II, "Message to the Pontifical Academy of Sciences: On Evolution," October 22, 1996.

7. Barbara Marx Hubbard, *Emergence: The Shift from Ego to Essence* (Charlottesville, VA: Hampton Roads Publishing Company, Inc., 2001), 4.

8. Mary Conrow Coelho, *Awakening Universe, Emerging Personhood: The Power of Contemplation in an Evolving Universe* (Lima, OH: Wyndham Hall Press, 2002) 39–41.

9. Caroline Webb, "Weaving a World with Light," *Earthlight,* 14, 1(Spring 2004), 27.

CHAPTER SIX: I AM THE LIGHT OF THE WORLD: INCARNATION AND PHOTOSYNTHESIS

1. Denis Edwards, *Jesus and the Cosmos* (Mahwah, NJ: Paulist Press, 1991), 69.

2. To say that Jesus is definitive is not to say that he is exclusive. So much harm has occurred as religious factions fight over the question about who has the "real" God. The answer: We all do. God is not confined by religion, only human beings are.

CHAPTER SEVEN: YOU ARE THE LIGHT OF THE WORLD: MORPHOGENIC FIELDS

1. Denis Edwards, *op. cit.,* 69.

2. Thomas Aquinas, *In Eph.* 3, lectio 5. In Lamb, *Aquinas Commentary,* 147; in Matthew Fox, *The Coming of the Cosmic Christ* (San Francisco: Harper and Row, 1988), 116.

3. Bede Griffiths, *A New Vision of Reality: Western Science, Eastern Mysticism and Christian Faith* (Springfield, IL: Templegate Publishers, 1989), 20.

4. Rupert Sheldrake, "In the Presence of the Past: An Interview with Rupert Sheldrake," http://twm.co.nz/shel -int2.htm.

5. Bede Griffiths, op. cit., 22.

6. Thomm Hartman, "Morphic Resonance," www.earca ndles.co.uk/uk/articles/article_morphicreonance.htm.

7. The PEAR website contains more detailed information, including charts of the statistical analysis: wttp://noosphere.princeton.edu. In addition, an article by Peter von Buengner, "Morphic Fields Can Now Be Measured Scientifically!", gives an overview of the project and its findings: www.experiencefestival.com/a/Morphic_fields/id/10320.

8. "Who Is Roger Nelson?" Parapsychological Association website, www.parapsych.org.

9. von Buengner, "Morphic Fields."

10. Institute of Science, Technology and Public Policy. www.istpp.org/crime_prevention.

11. Clarissa Pinkola Estes, Ph.D. "Mis Estimados: Do Not Lose Heart," 2003. www.creativeresistance.ca/strength/do-not-lose-heart-clarissa-pinkola-estes.htm

CHAPTER EIGHT: FOR FREEDOM CHRIST SET US FREE: THE THEORY OF HOLONS

1. Ken Wilber, *A Theory of Everything* (Boston: Shambhala Publications, Inc., 2000), 40.

CHAPTER NINE: SELF-DISSOLUTION AND BLACK HOLES

1. Richard Wolfson, "Einstein's Relativity and the Quantum Revolution: Modern Physics for Non-Scientists," second edition, video course. (Chantilly, VA: The Teaching Company, 2000).

2. Laura A. Whitlock, Kara C. Granger, and Jane D. Mahon, "The Anatomy of Black Holes," wttp://imagine.gsfc.nasa.gov, 3.

3. Ibid.

4. F. L. Cross and E. A. Livingstone, *The Oxford Dictionary of the Christian Church* (Oxford: Oxford University Press, 1974), 1278.

CHAPTER TEN: DEATH AND SUPERNOVAS

1. Laura A. Whitlock, Kara C. Granger, and Jane D. Mahon, op. cit.

2. http://encarta.msn.com/encyclopedia_1741500695/Supernova.html. Hubble images of the Large Magellanic Cloud may be accessed through NASA's Astronomy Picture of the Day.

CHAPTER ELEVEN: MYSTERY:
DARK ENERGY AND DARK MATTER

1. Redshift refers to the measurement of light waves. Light waves moving toward us shift to the blue end of the light spectrum. Light waves moving away shift toward the red end of the light spectrum, hence the term redshift.
2. Sean M. Carroll, "Cosmology Primer," http://pancake. uchicago.edu/~carroll/cfcp/primer. Carroll is assistant professor in the Physics Department of the University of Chicago, the Enrico Fermi Institute, and the Kavli Institute for Cosmological Physics.
3. Sara Goudarzi, "Nearby Evidence for Dark Energy," Special to Space.com (Mar. 2005), www.space.com/science astronomy/050322_dark_energy.html.
4. Ibid.
5. Paul Preuss, "Dark Energy Fills the Cosmos," Science Beat, an online publication of Berkeley Lab, www.lbl.gov/Science-Articles/Archive/dark-energy.html.
6. John of the Cross, "The Dark Night," *The Collected Works of St. John of the Cross*, Kieran Kavanaugh, O.C.D. and Otilio Rodriguez, O.C.D. trans. (Washington, D.C.: Institute of Carmelite Studies, 1979), 295.
7. Daniel L. Schutte, "Holy Darkness," (OCP Publications, 1992).

CHAPTER TWELVE: THAT ALL MAY BE ONE: LIVING IN RADICAL AMAZEMENT

1. Matthew Fox, *The Coming of the Cosmic Christ* (New York: Harper and Row, 1988), 19.
2. Abraham Heschel, op cit., 74–75.

BIBLIOGRAPHY
The New Universe Story
and Christian Spirituality

"Astronomy Picture of the Day," NASA,
http://antwrp.gsfc.nasa.gov/apod/astropix.html

Baez, John and Ilja Schmelzer. "Hawking Radiation,"
http://math.ucr.edu/home/breeze/physics/Relativity/BlackHol
es/hawking.html.

Berry, Thomas. *The Dream of the Universe*. San Francisco: Sierra Club
Books, 1990.

_____. *The Great Work*. New York: Bell Tower, 1999.

Brewi, Janice and Anne Brennan. *Celebrate Mid-Life: Jungian Archetypes
and Mid-Life Spirituality*. New York: Crossroad, 1988.

Carroll, Sean. "Cosmology Primer,"
http://pancake.uchicago.edu/~carroll/cfcp/primer/index.html.

_____. "The Preposterous Universe,"
http://pancake.uchicago.edu/~carroll/preposterous.html.

Coehlo, Mary Conrow. *Awakening Universe, Emerging Personhood: The
Power of Contemplation in an Evolving Universe*. Lima, OH:
Wyndham Hall Press, 2002.

Cowen, Ron. "Galaxy Hunters: The Search for Cosmic Dawn," *National
Geographic* 203, no. 2 (February 2003): 2–29.

Cross, F. L. and E. A. Livingstone. *The Oxford Dictionary of the Christian
Church*. Oxford, England: Oxford University Press, 1974.

Dickinson, Terence, ed., "Astronomy 2003" calendar. Buffalo, NY:
Firefly Books, 2003.

Edwards, Denis. *Jesus and the Cosmos*. Mahwah, NJ: Paulist Press, 1991.

Estes, Clarissa Pinkola, Ph.D. "Mis Estimados: Do Not Lose Heart,"
2003. www.creativeresistance.ca/strength/do-not-lose-heart
-clariss-pinkola-estes.htm.

Finley, James. *The Contemplative Heart*. Notre Dame, IN: Sorin Books,
2000.

_____. *Christian Meditation*. San Francisco: HarperSanFrancisco,
2004.

Fox, Matthew. *The Coming of the Cosmic Christ*. San Francisco: Harper
and Row, 1988.

Fox, Matthew and Rupert Sheldrake. *Natural Grace: Dialogues on Creation, Darkness and the Soul in Spirituality and Science*. New York: Doubleday Image Books, 1997.

Gilby, Thomas. *St. Thomas Aquinas: Theological Texts*. Durham, England: Labyrinth Press, 1982.

Goldman, Steven L. "Science in the Twentieth Century: A Social-Intellectual Survey." Chantilly, VA: The Teaching Company, 2004.

Goudarzi, Sara. "Nearby Evidence for Dark Energy," www.space.com/scienceastronomy/050322_dark_energy.html.

Griffiths, Bede. *A New Vision of Reality: Western Science, Eastern Mysticism and Christian Faith*. Springfield, IL: Templegate Publishers, 1989.

Guiley, Rosemary Ellen. *The Quotable Saint*. New York: Checkmark Books, 2002.

Gunderson, P. Erik. *The Handy Physics Answer Book*. Canton, MI: Visible Ink Press, 1999.

Hart, Thomas. *Spiritual Quest: A Guide to the Changing Landscape*. Mahwah, NJ: Paulist Press, 1999.

Hartman, Thomm, "Morphic Resonance," www.earcandles.co.uk/uk/articles/article _ morphicresonance.htm.

Hawking, Stephen. *A Brief History of Time*. New York: Bantam Books, 1988.

Heschel, Abraham Joshua. *God in Search of Man: A Philosophy of Judaism*. New York: Farrar, Straus and Giroux, 1955.

Hodgson, Peter, *Winds of the Spirit: A Constructive Christian Theology*. Minneapolis: Fortress Press, 1994, in Diarmuid O'Murchu, *Evolutionary Faith: Rediscovering God in Our Great Story*. Maryknoll, NY: Orbis Books, 2002.

Hubbard, Barbara Marx. *Conscious Evolution: Awakening the Power of Our Social Potential*. Novato, CA: New World Library, 1998.

_____. *Emergence: The Shift from Ego to Essence*. Charlottesville, VA: Hampton Roads Publishing Company, 2001.

John Paul II. "Message to the Pontifical Academy of Sciences on Evolution," October 22, 1996.

Johnson, Elizabeth, "Revival of the Cosmos in Theology," www.catholic-church.org/canossianssg/Ministries/justice/ revival_of_the_cosmos_in_theolog.htm.

_____. *Women, Earth, and Creator Spirit*. 1993 Madeleva Lecture in Spirituality. Mahwah, NJ: Paulist Press, 1993.

Johnston, William. *Arise, My Love: Mysticism for a New Era*, Maryknoll, NY: Orbis Books, 2000.

————. *Mystical Theology: The Science of Love*. Maryknoll, NY: Orbis Books, 1995.

Joseph, Lawrence E. "James Lovelock, Gaia's Grand Old Man," http://dir.salon.com/people/feature/2000/08/17/lovelock/index.html.

Kavanaugh, Kieran, O.C.D. and Otilio Rodriguez, O.C.D., trans. *The Collected Works of St. John of the Cross*. Washington, D.C.: The Institute of Carmelite Studies, 1979.

Keating, Thomas. "The Contemplative Journey," Audiotape series, Boulder, CO: Sounds True.

Keck, L. Robert. *Sacred Eyes*. Boulder, CO: Synergy Associates, Inc., 1992.

Livio, Mario. *The Accelerating Universe*. New York: John Wiley and Sons, Inc., 2000.

Lovelock, James. *Gaia: A New Look at Life on Earth*. New York: Oxford University Press, 1995.

Marshall, Ian and Danah Zohar. *Who's Afraid of Schrödinger's Cat?: An A-to-Z Guide to All the New Science Ideas You Need to Keep Up with the New Thinking*. New York: William Morrow Publishing, 1997.

May, Gerald G. *The Dark Night of the Soul: A Psychiatrist Explores the Connection Between Darkness and Spiritual Growth*. San Francisco: HarperSanFrancisco, 2004.

O'Murchu, Diarmuid. *Evolutionary Faith: Rediscovering God in Our Great Story*. Maryknoll, NY: Orbis Books, 2002.

————. *Quantum Theology: Spiritual Implications of the New Physics*. New York: Crossroad Publishing, 1997.

————. *Reclaiming Spirituality*. New York: Crossroad Publishing Company, 1997.

Preuss, Paul. "Dark Energy Fills the Cosmos," www.lbl.gov/Science-Articles/Archive/dark-energy.html.

Puhls, Louis J., S.J. *The Spiritual Exercises of St. Ignatius*. Chicago, IL: Loyola University Press, 1951.

Quammen, David. "The Evidence for Evolution Is Overwhelming," *National Geographic* 206, no.5 (November 2004): 4–31.

Ray, Paul H. and Sherry Ruth Anderson. *Cultural Creatives: How 50 Million People are Changing the World*. New York: Three Rivers Press, 2000.

Rupp, Joyce. *The Cosmic Dance: An Invitation to Experience Oneness.* Maryknoll, NY: Orbis Books, 2002.

Schroeder, Gerald L. *Genesis and the Big Bang: The Discovery of Harmony Between Modern Science and the Bible.* New York: Bantam Books, 1990.

_____. *The Science of God: The Convergence of Scientific and Biblical Wisdom.* New York: Broadway Books, 1997.

Schutte, Daniel L. "Holy Darkness," OCP Publications, 1992.

Sheldrake, Rupert. "In the Presence of the Past: An Interview with Rupert Sheldrake," http://twm.co.nz/shel-int2.html.

Sobosan, Jeffrey G. *Romancing the Universe: Theology, Science, and Cosmology.* Grand Rapids, MI: William B. Eerdman's Publishing Company, 1999.

Soelle, Dorothee. *The Silent Cry: Mysticism and Resistance.* Minneapolis: Fortress Press, 2001.

"Supernova," http://encarta.msn.com.encyclopedia_1741500695/Supernova.html.

Swimme, Brian. *Canticle to the Cosmos* (video cassettes or DVD).

_____. *The Hidden Heart of the Cosmos: Humanity and the New Story.* Maryknoll, NY: Orbis Books, 1996.

_____. *The Powers of the Universe.* DVD available at brianswimme.org

_____. *The Universe Is a Green Dragon: A Cosmic Creation Story.* Santa Fe, NM: Bear and Company, 1984.

Swimme, Brian and Thomas Berry. *The Universe Story.* San Francisco: HarperSanFrancisco, 1992.

Snyder, Mary Hembrow, ed. *Spiritual Questions for the Twenty-First Century.* Maryknoll, NY: Orbis Books, 2001

Teasdale, Wayne. "The Interspiritual Age: Practical Mysticism for the Third Millennium," Council on Spiritual Practices. www.csp.org/experience/docs/teasdale-interspiritual.html.

_____. *The Mystic Heart: Discovering a Universal Spirituality in the World's Religions.* Novato, CA: New World Library, 1999.

Teilhard de Chardin, Pierre. *The Human Phenomenon.* Translated by Sarah Appleton-Weber. Portland, OR: Sussex Academic Press, 2003.

Vermaas, Wim. "An Introduction to Photosynthesis and Its Application," Arizona State University: http://photoscience.la.asu.edu/photosyn/education/photointor .html.

von Bruegner, Peter. "Morphic Fields Can Now Be Measured Scientifically!" www.experiencefestival.com/a/Morphic_fields/id/10320.

Webb, Caroline "Weaving a World with Light," *Earthlight* (*light* 14, no. 1 (Spring 2004): 27.

Wessels, Cletus. *The Holy Web: Church and the New Universe Story.* Maryknoll, NY: Orbis Books, 2000.

_____. *Jesus in the New Universe Story.* Maryknoll, NY: Orbis Books, 2003.

Whitlock, Dr. Laura A., Kara C. Granger and Jane D. Mahon, "The Anatomy of Black Holes," *Imagine the Universe* series, http://imagine.gsfc.nasa.gov.

"Who Is Roger Nelson?", Parapsychological Association, www.parapsych.org.

Wilber, Ken. *A Brief History of Everything.* Boston, MA: Shambala Publications, 1996.

_____. *Kosmic Consciousness.* Audio CDs. Boulder, CO: Sounds True, 2003.

_____. *Sex, Ecology, and Spirituality: The Spirit of Evolution.* Boston, MA: Shambala Publications, 2000.

_____. *A Theory of Everything.* Boston, MA: Shambala Publications, 2000.

Wolfson, Richard. "Einstein's Relativity" and the Quantum Revolution: Modern Physics for Non-Scientists," second edition video course, Chantilly, VA: The Teaching Company, 2000.

Zohar, Danah and Ian Marshall. *The Quantum Society: Physics and a New Social Vision.* New York: William Morrow and Company, 1994.

_____. *SQ: Connecting with Our Spiritual Intelligence.* New York: Bloomsbury Publishing, 2000.

Zohar, Danah. *The Quantum Self: Human Nature and Consciousness Defined by the New Physics.* New York: William Morrow Publishing, 1990.

JUDY CANNATO is a spiritual director at River's Edge: A Place for Reflection and Action at St. Joseph Center in Cleveland, Ohio, as well as a retreat director, certified Mid-Life Directions Consultant, and associate member of the Congregation of St. Joseph. She has master's degrees in both Education and Religious Studies from John Carroll University, where she was a member of the adjunct faculty in the Department of Religious Studies. She is the author of *Quantum Grace: Lenten Reflections on Creation and Connectedness* and *Quantum Grace: The Sunday Readings*, as well as numerous articles on spiritual life. Judy and her husband have two grown sons.

Other Titles by Judy Cannato

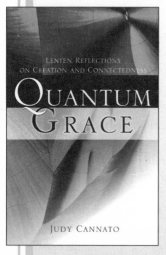

QUANTUM GRACE
Lenten Reflections on Creation and Connectedness

Invites us, during Lent, to look at our beliefs, decisions and actions, and the way they affect our personal lives and the lives of others. Intended for daily individual reflection, but also ideal for group use.
ISBN: 0-87793-984-5 / 160 pages / $9.95

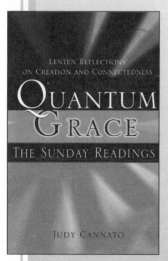

QUANTUM GRACE
The Sunday Readings

Quantum Grace: The Sunday Readings expands on the author's theme of observing Lent through the lens of "new physics." It seeks to stir the reader to look at Scripture through a slightly different lens—one that reconciles our Christian tradition with modern scientific discoveries.
ISBN: 1-59471-024-4 / 128 pages / $9.95